The Politics of Military Coalitions

Military coalitions are ubiquitous. The United States builds them regularly, yet they are associated with the largest, most destructive, and consequential wars in history. When do states build them, and how do they choose partners? Are coalitions a recipe for war, or can they facilitate peace? Finally, when do coalitions affect the expansion of conflict beyond its original participants? *The Politics of Military Coalitions* introduces newly collected data designed to answer these very questions, showing that coalitions – expensive to build but attractive from a military standpoint – are very often more (if sometimes less) than the sum of their parts, at times encouraging war while discouraging it at others, at times touching off wider wars while at others keeping their targets isolated. The combination of new data, new formal theories, and new quantitative analysis will be of interest to scholars, students, and policymakers alike.

Scott Wolford is an associate professor of government at the University of Texas. He has published articles in *The American Journal of Political Science*, *Journal of Politics*, *Journal of Conflict Resolution*, and *International Studies Quarterly*, among others. He is a Fellow of the Frank C. Irwin Chair in Government (2011–2016) and a recipient of the Best Paper in International Relations Award from the Midwest Political Science Association (2009).

The Politics of Military Coalitions

SCOTT WOLFORD

University of Texas at Austin

CAMBRIDGE
UNIVERSITY PRESS

CAMBRIDGE
UNIVERSITY PRESS

32 Avenue of the Americas, New York NY 10013-2473, USA

Cambridge University Press is part of the University of Cambridge.

It furthers the University's mission by disseminating knowledge in the pursuit of education, learning and research at the highest international levels of excellence.

www.cambridge.org
Information on this title: www.cambridge.org/9781107496705

© Scott Wolford 2015

First published 2015
First paperback edition 2016

A catalogue record for this publication is available from the British Library

ISBN 978-1-107-10065-7 Hardback
ISBN 978-1-107-49670-5 Paperback

Contents

List of Figures

List of Tables

Preface

I honestly never intended to write this book – or any book, for that matter. My plan had always been a career based on journal articles, which is pretty much the norm for someone like me who publishes applied formal theory and an occasional empirical paper. However, armed with a fellowship semester in the fall of 2012, some data I'd started collecting but didn't know what to do with, and a couple of related journal articles in the pipeline, the idea to tie it all together into a book started to take shape. What, after all, did we *really* know about military coalitions? Was the conventional wisdom, basically a horror story of free-riding, intramural rancor, and coercive ineffectiveness, correct? If it was, why did we see so many states using coalitions in so many conflicts? Did coalition partners increase the chances of war? Did they presage expanded conflicts? I soon realized that, to do this thing right – that is, to introduce my own definition of military coalitions, to build a series of theoretical models, and to test their predictions against the data – would be very much a book-sized endeavor.

After lamenting each of the (many, many) times I'd tuned out when people talked about book writing and publishing, I took a leap into the dark and got to work. The result, I hope, can start a scholarly dialogue about the dominant mode of military cooperation in international politics: the crisis-specific, purpose-built coalition. In the chapters that follow, I introduce the data, identify some puzzles to be explained, and then analyze the choice of partner, the escalation of crises to war, and the expansion of conflicts, all with an eye to the fundamental politics of buying and selling military cooperation. Hopefully, my claims that coalitions are worthy of study, that when it comes to shaping patterns of international war and peace they are more than the sum of their parts, are convincing. At the same time, I'm also hopeful that the exercise invites scrutiny and improvement; to the extent I can be productively wrong about something in this book, I'll consider the effort a success.

The effort, though, wasn't all my own; I racked up intellectual debts over the last several years like a compulsive gambler. First, I should acknowledge the people who set me most clearly on the path to academia: Jeff Freyman and Don Dugi at Transylvania University. Without their guidance, which involved a characterization of my writing as "like an 18th Century French philosopher" (not a compliment, mind you) and the (correct) observation that I needed to take some "pride in workmanship," I might've become . . . a lawyer. I spent my first year of graduate school at the University of Kentucky, where I studied under Doug Gibler and Matt Gabel, a brief time that still exercises a disproportionate influence on my work to this day. Finally, the seeds of this project were sown during my time at Emory University, where I learned from Dan Reiter, Cliff Carrubba, Eric Reinhardt, and Micheal Giles; all of them rightly picked up on the need to convince me to work a little bit harder, to put in that extra bit of spit and polish, and while I still struggle with that, I'm a better scholar for their efforts to push me in the right direction.

Countless people and organizations deserve mention for their encouragement, feedback, and support over the last two years of shepherding this book through the writing process: the University of Texas for my free semester in Fall 2012; the University of Colorado and the Colorado European Union Center of Excellence for the grant that allowed me to hire Amber Curtis to start data collection in 2009; audiences at SUNY Buffalo, the University of Rochester, the University of Virginia, and Texas A&M University who invited me to present parts of the project over the years; the students in my Fall 2013 International Security seminar, who read an early draft of the manuscript; people who offered helpful comments along the way, like Harrison Wagner, Amy Yuen, Kris Ramsay, Harvey Palmer, Ric Stoll, William Spaniel, Dan McCormack, Henry Pascoe, and Cliff Morgan; a brilliant set of dear friends who generously read full drafts of the book, in some cases multiple times: Emily Ritter, Toby Rider, Phil Arena, and my fantastic IR colleagues here at UT, Terry Chapman, Mike Findley, Pat McDonald, and Rachel Wellhausen; the editors and anonymous reviewers at the *American Journal of Political Science* and *International Studies Quarterly*, where I published parts of Chapter 4 and an early version of Chapter 5; the staffs at the Dog & Duck Pub and Draft Pick in Austin, Cerveteca Lisboa in Lisbon, and Club 93 in Budapest, for excellent writing environments; and, of course, three anonymous reviewers and my editor, Robert Dreesen, at Cambridge University Press, who made this process a smooth and, ultimately, a fun one. My family also deserves a lot of credit; my parents, Mike and Susanne, have been unwavering in the enthusiasm of their support, and my wife, Amy, has been loving, helpful, brilliant, supportive, and – as she's already been through the book-writing process herself – remarkably patient and understanding. I couldn't have done this without her.

1

Introduction

[M]uch the greater part of the study of the authoritative allocation of value is reduced to the study of coalitions.

William H. Riker
The Theory of Political Coalitions, p. 12

Military cooperation is ubiquitous in international politics. States have a long tradition of signing treaties of alliance for collective defense, like the North Atlantic Treaty Organization, as well as actively fighting wars together, like the coalitions that twice faced Napoleonic France in hopes of preserving the European balance of power. They have colluded to conquer and partition their neighbors, as provided for in the Molotov-Ribbentrop Pact signed between Nazi Germany and the Soviet Union, and they have built coalitions to threaten war lest targets change their policies, from the Boxer Rebellion in 1900 to the Iraqi annexation of Kuwait in 1990 to ethnic cleansing in Kosovo in 1999. In fact, fully 40% of interstate wars in the past two centuries have been multilateral (Sarkees and Wayman 2010), and while only about one-quarter of international crises since the Second World War have seen coalitions form on at least one side (Wolford 2014a), the United States has built coalitions of varying sizes to support nearly half of its own uses of force since 1948 (Tago 2007), and almost all since the end of the Cold War (Kreps 2011). Yet despite their pervasiveness, to say nothing of their popularity with contemporary great powers, we still know little about how military coalitions and the efforts taken to hold them together affect patterns of international war and peace.

The occurrence and expansion of armed conflict are of enduring interest to students of international relations, all the more so when states cooperate militarily: coalitions have been integral to some of the most destructive events in international history, and multilateral wars are among the longest, bloodiest, and widest-ranging in their implications.[1] In the twentieth century alone, two world wars redrew the global political map as rival coalitions fought on a nearly apocalyptic scale for the domination of continents and oceans, eliminating and creating both individual states and entire international orders and fundamentally altering the power relationships in the international system. In the following decades, bloody aftershocks in the divided countries of Korea and Vietnam, each born of rivalries forged in the disintegration of the Second World War's victorious coalition, saw the United States fight alongside new coalition partners to preserve the Cold War status quo against communist-led attempts to overturn it. Even conservative estimates suggest that these four wars killed nearly 30 million combatants, to say nothing of tens of millions more civilians, in a death toll that would scarcely have been possible had not states cooperated to wage war on one another.[2]

Nonetheless, wide-ranging systemic wars are the exception, not the rule. Coalitions and military cooperation are also prevalent in the smaller-scale conflicts that account for the vast majority of wars, as well as in the crises, tensions, and militarized disputes that precede them. Since the end of the Cold War in 1989, coalitions of varying sizes have participated in two American-led wars against Iraq (1991 and 2003), as well as multilateral interventions in the former Yugoslavia (1995, 1999), Libya (2011), and Iraq and Syria against the self-styled "Islamic State" (2014). Of course, coalitions do not simply fight wars. Before crises escalate to full-scale violence with armies clashing on the battlefield, fighters and bombers filling the skies, and navies facing off at sea, states cooperate to make collective threats of war, hoping to achieve their aims peacefully without having to make good on those threats. For every crisis that boils over into war, coalitions achieve their aims by coercing their desired concessions short of violence in numerous others, just as the vast majority of coalitions manage to keep their conflicts localized rather than precipitate globe-spanning conflagrations. Finally, acting with "friends and allies" is often not only useful militarily but also good politics at home and abroad (Chapman 2011, Chapman and Reiter 2004, Lai and Reiter 2005, Nye 2002). As such, in the post–Cold War era, multilateral action has become the "default" for American foreign policy (Kreps 2011), whether it involves active cooperation in the

---

[1] See Slantchev (2004) and Shirkey (2012) on the link between the number of actors and war duration and severity, respectively.

[2] Total battle deaths across these four wars as calculated by the Correlates of War project (Sarkees and Wayman 2010) are 27,144,464, but this total omits the civilian toll in each conflict, as well as deaths in related conflicts: the Japanese invasion of China, the Chinese Civil War, the Nomonhan campaign, and the Indochina War. Hastings (2012, p. 646), for example, reports 60 million deaths, soldiers and civilians alike, as a conservative estimate for the Second World War alone, and recent estimates indicate that military deaths in the First World War have also been underestimated, placing the updated total around 10 million (Prost 2014).

application of military force, the formal support of international institutions, or both.

For all their ubiquity, it should not be surprising that coalitions produce a diverse set of political outcomes, from successful – that is, peaceful – coercion to the outbreak of war, and from localized conflicts against isolated targets to wide-ranging confrontations that draw neighbors and distant powers alike into counter-coalitions. These patterns raise two obvious questions. First, how do military coalitions shape the probability of war and the prospects for peace? Next, when do coalitions provoke counter-coalitions and the expansion of conflicts? To answer these questions, I argue that we must begin with a prior understanding of when and with whom states choose to build military coalitions in the first place, especially the means by which such cooperation is negotiated, secured, and preserved. By understanding *how* states negotiate the terms of and secure military cooperation, we can develop a better understanding of the consequences of that cooperation for patterns of international war and peace.

Joining a coalition is inherently costly, from the upfront expenses of mobilization and war to the opportunity costs of abandoning more immediate priorities. Quite apart from divergent assessments of the value of the prize for which they fight, coalitions also disagree internally over the distribution of the costs and risks of war, which inevitably fall differently across their members. Even on the Western Front in the First World War, British and French national priorities led to intramural clashes over the distribution of effort, territory, and casualties (Hastings 2013, Herwig 2011, Philpott 2014); as a leading scholar of the period puts it, "Clausewitz's famous dictum about politics and war applies in struggles between friends as well as in conflicts with enemies" (Philpott 2011, p. 235). The costs of cooperation ensure that partners must be compensated in return for their assistance. Cooperation is transactional, and settling on the terms of working together, or the price of cooperation, is often the primary challenge faced by would-be coalition partners. This compensation may be direct side payments, influence over the spoils of victory, indemnities and reparations, or compromises over military and bargaining strategies – none of which a coalition-builder would like to yield or change if it can avoid doing so. In the following chapters, I show that such compensation, whatever form it takes, can have profound implications for both whether states cooperate *and* how they go on to conduct coercive diplomacy against their enemies. In other words, the terms of cooperation that facilitate the construction and maintenance of military coalitions can then go on to have second- and third-order effects on the probability of war and the expansion of conflicts to include other belligerents.

This book addresses three intimately related problems using a combination of theoretical and empirical models, as well as newly coded data on coalitions in international crises. First, and fundamentally, it outlines the conditions under which states build coalitions in the shadow of war and what partners they choose, explaining cooperation not solely in terms of shared interests but also in terms of how compensation can be used to overcome divergent interests. Second, it explores how the choice of partner affects the escalation of disputes to war, in particular how preserving cooperation among one's own coalition partners affects the processes of making military threats and signaling resolve to an opponent that doubts a coalition-builder's willingness to fight. Finally, it analyzes how the prior choice of partners affects the expansion of conflicts to draw in other states, identifying when coalition-building touches off balancing responses and the formation of counter-coalitions, as well as the conditions under which victorious coalitions disintegrate and come to blows over the terms of the peace just secured. Throughout, I trace the effects of two crucial factors – military power and foreign policy preferences – on each stage of this process, showing how they can advance our understanding of the role of coalitions and military cooperation in the patterns of war and peace that define international relations. I also use the logic of the theory to shed light on the success (and failure) of American-Turkish coalition negotiations in 1991 and 2003, the signaling challenges facing the United States in the 1961 Berlin Crisis and the 1999 Kosovo War, as well as the responses of third parties to the two American wars against Iraq and the Soviet invasion of Afghanistan in 1979.

1.1 RETHINKING MULTILATERALISM

At its core, this is a book about international cooperation. However, where the majority of work on the topic explores the role of formal institutions such as alliances or international organizations in preserving peace, eliminating trade barriers, or resolving policy disputes (e.g., Gowa and Mansfield 1993, Keohane 1984, Morrow 2000, Oye 1986, Rosendorff 2005, Russett, Oneal, and Davis 1998), my focus is on cooperation in the form of military coalitions, purpose-built for an ongoing or imminent crisis, which may be formal *or* informal. Cooperation on the home front, of course, is a requisite for waging war; armies must be raised, supplied, fielded, and commanded, and success at each stage of this process requires that individuals, often very large numbers of them, overcome numerous, potentially severe collective action problems (Wagner 2007, ch. 3,4). Yet cooperation does not stop once states have put the economy on a war footing, raised armies, trained pilots, and put ships to sea; they often aggregate their military power by making threats of collective military action or, when those threats fail, by fighting in common cause to compel their opponents to do their will. As I argue later in the book, whether and how they secure this military cooperation can have wide-ranging – and often surprising – implications for patterns of international war and peace.

Cooperation between states is the result of "mutual adjustment" (Keohane 1984, p. 13), whereby states coordinate erstwhile-divergent policies, taking actions they otherwise would not, in order to achieve some common goal (*ibid.*, p. 52).[3] While cooperation is costly, it produces goods that would be difficult or impossible to produce in its absence. Traditional approaches to international relations tell a number of different stories about international cooperation, particularly the military variety. Some structural realists are famously skeptical of the feasibility of international cooperation in general and balancing coalitions in particular (Mearsheimer 2001, p. 212), even as others assert that states tend to cooperate in the recurrent formation of balancing coalitions against the powerful (Waltz 1979) or threatening (Walt 1987).[4] On the other hand, states often remain neutral in the face of rising threats, staying on the sidelines and hoping to shift the burden of preserving the balance of power on to other like-minded states (Christensen and Snyder 1990, Powell 1999), a compelling indication of the costs entailed in cooperating militarily alongside coalition partners. Nonetheless, as attested to by the emergence of coalitions in interstate crises and wars, states often *do* cooperate militarily for a variety of reasons, only some of which are related to rising threats to international order. Understanding the origins, politics, and consequences of the "diplomacy of co-belligerency" (Fowler 1969, p. 4), ubiquitous in its frequency and oftentimes infamous in its consequences, is the central theme of this book.

In the chapters that follow, I make the case for both refining and expanding the scope of the analysis of military cooperation. I refine the traditional mode of analysis by choosing a deliberately narrow definition of military coalitions, one centered on cooperative actions in discrete international crises. At the same time, this narrower unit of observation expands on the traditional empirical scope, allowing for a consideration of both allied and nonallied forms of cooperation, as well as states other than great powers. Thus, rather than identifying (a) opportunities to balance, bandwagon, or pass the buck in the face of threats to the balance of power, as do many structural realist accounts (Christensen and Snyder 1990, Mearsheimer 2001, Walt 1987, Waltz 1979), or (b) formal promises to fight together in treaties of alliance, as is the case in most quantitative work (Leeds 2003a, 2003b), I examine choices over coalition formation in the context of international crises. In other words, I analyze strategic decisions to cooperate in the context of ongoing crises, where the participants need not (but may) be allied, and where the issues at stake need not (but may) be tied to the processes of great power politics.

Focusing on the international crisis – a period of heightened tension in which general deterrence has broken down and where war is possible, but not inevitable, between two or more states (Wilkenfeld and Brecher 2010) – offers several advantages for the study of military cooperation. First, though

---

[3] Keohane distinguishes cooperation from a situation of harmony, in which states' policies serve each other's mutual purposes with no adjustment – a situation both uninteresting and, especially when it comes to cooperating in the costly endeavor of interstate war, highly unlikely.

[4] They may also cooperate by bandwagoning, hoping to accommodate themselves to the powerful or share in the spoils of victory (Schweller 1994).

of smaller scale than great power conflicts and system-ordering general wars, crises are also far more common, making for a larger sample of opportunities for cooperation and, as a result, potentially stronger inferences. Second, despite the diversity of issues over which they arise, international crises are structurally similar events, in that each participant in a crisis confronts a common series of questions: How much help do I need? How much am I willing to pay in return for it? And is there anyone out there willing to help for an acceptable price? Thus, a crisis-specific analytical focus brings into sharper relief the basic transactional nature of international cooperation and coalition-building, as well as the onset of war and the expansion of conflicts. Third, as I discuss at greater length in Chapter 2, the approach has empirical advantages, as the crises that drive states to build coalitions in the shadow of war are useful for identifying a large number of discrete cases in which cooperation and its absence, not just in terms of taking the same side militarily but also concessions and policy adjustments, are easy to define and identify.

More broadly, this book helps clarify the distinction between two types of multilateralism: military and diplomatic. While a large body of research – to say nothing of popular discourse – collects only the latter (Chapman 2011, Nye 2002, Thompson 2006, Voeten 2005) or some combination of the two (Kreps 2011, Tago 2007) under the single heading of "multilateralism," my focus is squarely on the military variety, where states contribute materially (or threaten to do so) in a cooperative application of military force. As I argue in Chapter 2, the two processes are interrelated, though analytically and empirically distinct. Diplomatic multilateralism, which derives from the sanction or approval of international institutions, can facilitate military cooperation, easing fears of expansionism in states that might otherwise refuse cooperation (Voeten 2005). However, extant scholarship that focuses on diplomatic multilateralism does not explain why states choose the military partners they do – if they choose any at all – when faced with an international crisis, either with or without the support of international institutions. Rather, developing an understanding of partner choice, as well as the subsequent crisis behaviors linked to it, requires an understanding of the processes of bargaining, compensation, and cooperation between coalition members that I develop in Chapter 3.

Given the gulf between their potentially dire consequences and their prominence in the historical record, coalitions and military cooperation should play a large role in how we answer some of the most enduring questions about war and peace in international politics. Thus far, this role has been too small. As I argue in Chapter 2, there is a paucity of answers to the puzzles of cooperation, war, and conflict expansion that can apply directly to the context of coalitions and military cooperation. Such answers as we have are not based on a strong theoretical foundation that reflects the transactional nature of military cooperation, and this poses significant obstacles to a useful understanding of the role of military coalitions in everyday international politics. Advancing our understanding of the role of military cooperation in international relations, I argue, requires a unified theoretical and empirical approach that integrates a theory of cooperation, one that acknowledges the centrality of securing and preserving costly military cooperation, with complementary theories of both (a) crisis bargaining and war and (b) alignment decisions and conflict expansion.

1.2 THE ARGUMENT

The book's primary argument is that untangling the relationships between military cooperation, war, and conflict expansion requires an understanding of the key political dynamics that define any coalition: the negotiation and maintenance of military cooperation. This requires new theoretical models, combining the underlying transactional theory of military cooperation with models of coalition formation, crisis bargaining, and alignment decisions, as well as a unique empirical strategy that introduces new data and units of analysis. At the center of the story is the notion that states choose coalition partners by weighing the expected costs of securing and maintaining their cooperation against the potential military benefits of acting multilaterally, while their partners weigh the benefits of proposed compensation against the costs of joining and, potentially, fighting alongside other states as part of a coalition. As a result, the concessions or compromises made to ensure cooperation, particularly when they affect crisis bargaining strategies and the durability of coalitions, can have implications for the escalation of crises to war and the expansion of ongoing conflicts.

As discussed in Chapter 2, coalitions are crisis-specific phenomena. They exist when states make threats of collective military action – generally, war – against a target state lest it change its behavior or make some desired policy concession. Thus, whether coalition partners ultimately fight together depends first and foremost on whether their targets reject their demands. This requires defining "coalition" as an autonomous concept, distinct from other forms of military cooperation. Coalitions are not alliances, incomplete contracts concerning behavior in war that may or may not be activated or honored in a given crisis (Benson 2012, Morrow 2000), nor are they instances of diplomatic multilateralism, where other states or institutions offer political support but need not participate militarily (see Kreps 2011). Rather, military coalitions involve short-run, crisis-specific decisions over military cooperation for which treaties of alliance and widespread diplomatic support are neither necessary nor sufficient. In addition to fleshing out this concept of coalitions, Chapter 2 also introduces a new dataset of coalitions and their participation in international crises on which the empirical models of subsequent chapters are based, then conducts some preliminary empirical analyses to determine whether and how coalitions differ from states acting alone. In particular, it shows that coalitions see higher rates of both crisis escalation and expansion than do states acting unilaterally, but a pair of decomposition analyses also shows that these differences cannot be explained solely with reference to their uniquely high levels of military power. Coalitions, in other words, cannot be usefully understood as mere aggregations of power.

After Chapter 2 establishes the facts to be explained and the puzzles to be solved, the theoretical models of Chapters 3–5 show that two key variables, military power and diversity in foreign policy preferences, interact to shape outcomes across each stage of this process of military multilateralism: (a) coalition formation, (b) crisis escalation, and (c) conflict expansion. The last stage depends on expectations about a war-winning coalition's subsequent durability, which I also subject to empirical analysis. Fundamental to each stage of the process is the maintenance of costly military cooperation in the face of incentives to refuse or defect, a challenging goal made more difficult by the inherent diversity of preferences among states in the international system. States differ in their evaluations of the status quo, valuations of the stakes of a given conflict, tastes for risk, and willingness to bear costs, meaning that would-be coalition builders must compensate partners for participation in the inherently costly enterprises of crisis bargaining and military coercion. Throughout the process, the costs of securing military cooperation and the willingness of a lead state to pay those costs influence whether states cooperate to form coalitions, whether they can achieve their goals peacefully, and whether their conflicts expand beyond their initial participants.

To develop the theory, I begin with a highly stylized model of coalition formation and crisis initiation built around the basic insight that securing costly military cooperation in crises requires compensation. Then, I analyze increasingly complicated models of crisis escalation and expansion, exploring the challenges of military cooperation in the context of the theory's two primary concepts – the distribution of preferences within the coalition and its aggregated or relative military power. The first pairing of theoretical and empirical models, presented in Chapter 3, focuses on coalition formation and shows that states in a crisis face a trade-off between enhancing their military prospects and making costly concessions to ensure a potential partner's cooperation. While states prefer partners with preferences similar to their own, they become decreasingly selective the more a given partner enhances their military prospects, building coalitions around increasingly diverse sets of preferences – a diversity that will go on to have some surprising effects on both crisis escalation and conflict expansion.

The models in Chapter 4 integrate the process of coalition formation and compensation with crisis bargaining and costly signaling, allowing me to explore how the formation of diverse coalitions affects the ability to send credible signals of resolve.[5] The key insight is that the costs of war may fall unequally across coalition members, which makes some potential partners hesitant about participating in overly costly wars. When a lead state risks losing a

---

[5] Chapter 4 uses the theoretical model of Wolford (2014b), but the empirical model and case discussions are unique to this book.

partner's cooperation if it threatens an intolerably costly war, it chooses escalation strategies that weigh the demonstration of resolve against the maintenance of military cooperation. I show that, when targets are strong, irresolute coalition leaders will be deterred from bluffing, preserving both military cooperation within the coalition and peace with its opponent. However, when targets are relatively weak, resolute coalition leaders may knowingly mask their resolve, preserving a partner's cooperation despite the fact that leaving the opponent's information problem unsolved raises the chances of war. Therefore, the presence of coalition partners can encourage either peace or war, depending on how their desire to restrain the lead state affects threat-making and signaling decisions in crises; in fact, attempts to rein in the lead state can reduce the expected costs of war while simultaneously making war more likely.

Chapter 5 presents models that show how military power and preference diversity interact in the final stage of military multilateralism, affecting the expansion of conflicts through the provocation of balancing and the durability of victorious coalitions after wars.[6] Third-party states are frequently concerned about threats posed to them by the victors of ongoing conflicts (Powell 1999, Voeten 2005, Walt 1987), so to the extent that coalition members are more threatening together than apart, third parties are particularly concerned with whether a coalition involved in a conflict today will stay together and threaten additional states in the future. As I argue later, the calculations of future threat differ when a coalition – as opposed to a single threatening state – is involved. Balancing, however, is costly and risky, and third parties must weigh the chance to defeat or split coalitions against the uncertain threat posed by their members in the future. In a process unique to multilateral contexts, a coalition's revealed foreign policy preferences help outsiders form judgments about coalitional durability and, by extension, threats posed by coalitions in the future. Just as they did in the previous models, power and preferences again interact, such that an increasing diversity of coalitional preferences discourages the expansion of conflicts when coalitions are powerful. However, such diversity actually facilitates expansion when coalitions are weaker. This model also generates predictions over the durability of coalitions following victory in war, and I use additional data to show that, as predicted, homogeneous coalitions tend to maintain postwar cooperation longer than more diverse coalitions.

Chapter 6 closes the book with a summary of the project's contributions to the literatures on international conflict and cooperation, particularly how a focus on military cooperation can shed light on fundamental questions about military conflict, cooperation, and the sustainability of postwar peace. It also includes discussions of avenues for future research on the topic of military multilateralism, including how coalition-building differs by the relative share

---

[6] The theoretical model is a generalization of the game analyzed in Wolford (2014a), while the empirical models and case discusions expand substantially their counterparts in the original article.

of public and private goods at stake, the differences between allied and non-allied coalitions comma and a consideration of the theory's implications for American foreign policy. Taken together, the book's pairings of theoretical and empirical models paint a richer picture of military cooperation than existing theories allow, and each step of the modeling process builds on the former by leveraging a common set of concepts and assumptions. By tracing the effects of military power and preference diversity through the entire process of military multilateralism, it shows that the presence of coalition partners can either encourage or discourage both war and the expansion of conflicts. Coalitions are at times more prone to war than are states acting alone, at times less; at times better able to keep their conflicts localized, at times worse than states acting alone. They are, in a very real sense, more than the sum of their parts. Further, the approach adopted here makes these comparisons without first begging the question of why coalitions form in the first place, providing a more consistent theoretical account than extant work does to support the hypotheses that are derived and tested empirically in each chapter.

1.3 CONCLUSION

Questions of military cooperation, the outbreak of war, and the expansion of conflict sit at the core of international relations theory, because all three phenomena play enduring and constitutive roles in shaping the ebb and flow of what we recognize as international politics (Braumoeller 2012, Bull 1977, Wagner 2007). I have argued that scholarship too often pays short shrift to the role played by processes of military cooperation in both war and conflict expansion, and drawing out the links between coalitions, war, and the spread of military conflict is worthwhile in its own right. However, in addition to explaining specific patterns related to military cooperation, the book touches on several other important themes.

Most obviously, it highlights the relationship between concepts that too often sit on opposite sides of what is, in a very real sense, a false dichotomy: conflict and cooperation. Without cooperation, the armies with which wars are fought – indeed, the states that fight them – could scarcely exist (cf. Wagner 2007), and yet the study of international conflict very often abstracts away from the role of coalition partners, whether potential or actual, in explaining the outbreak of war. Most empirical studies of international conflict are strictly dyadic (see, e.g., Bennett and Stam 2004).[7] However, the approach taken here, which identifies coalitions by their participation in specific international crises, pushes our understanding of the link between cooperation and conflict several steps further. Ultimately, this book shows that "cooperation" is multifaceted

---

[7] To be fair, the potential for allied intervention does emerge in some treatments of dispute initiation (Leeds 2003b, Werner 2000).

and, in many cases, a double-edged sword, one that can help avert war or bring it about, and one that can provoke balancing coalitions or discourage them.

Next, and more practically, the study of coalitions should encourage scholars to rethink our everyday empirical and theoretical uses of several common concepts, perhaps most notably military power. Standard dyadic measures of relative power are problematic in their own right when they omit partner contributions (see also Croco and Teo 2005, Poast 2010, Sobek and Clare 2013), but subsequent chapters show that merely accounting for total relative power can be distorting: changes in relative power can mean radically different things – war or peace, conflict expansion or localization – depending on how the diversity of preferences within coalitions affects signaling strategies and third party threat assessments. Thus, scholars should think hard about the link between theory and research design in the study of international war, in particular what their theoretical constructs mean for choices over units of observation and units of analysis, because coalitions are not the behavioral equivalent of powerful states; as I show in the following chapters, the negotiations required to build and maintain them ensure that they are more than the sum of their parts, altering bargaining strategies and intervention decisions in ways unanticipated, and unaccounted for, by the traditional focus on the dyad in international conflict. Coalitions often behave in ways and produce political outcomes radically different than do states acting alone, and the approach developed here offers a few first steps along the path to making these comparisons.

Finally, my hope is that, by taking some early steps toward thinking about military coalitions as a unique, and uniquely important, manifestation of international cooperation, the book proves fertile ground for future research. By identifying common variables and mechanisms across multiple empirical contexts, this book is an attempt to start to a "modeling dialogue" (Myerson 1992, Powell 1999) between theory and evidence. It uses newly collected data to identify a set of facts to be explained, proposes a set of first principles to characterize the strategic setting and on which to build theories, and offers an initial set of explanatory factors that future work may use to refine (or, even more productively, refute) the claims and arguments advanced throughout this text. As I discuss in the conclusion, the differences between allied and nonallied coalitions, the postwar durability of victorious coalitions, and the possibility of military cooperation in light of the distinction between public and private goods provide opportunities to expand this research in new and productive directions.

2

Why Coalitions?

> Even the notion of "coalition" does not have a clear-cut empirical ana-
> logue . . . [T]he existence of a coalition does not depend on formal agreement.
>
> Glenn H. Snyder (1991, p. 132)

In this chapter, I distinguish military coalitions from other, more commonly
studied forms of cooperation and show how they can provide new answers to
enduring questions about (a) the conditions under which states cooperate, (b)
the probability of war, and (c) the expansion of conflicts. A military coalition
exists when two or more states threaten another with collective military action
in an international crisis; that is, they promise to cooperate in the imposition
of military costs on a target state unless their demands are satisfied. Coalitions
formed in about one-quarter of interstate crises from 1946 to 2001 (Wilkenfeld
and Brecher 2010) as well as 40% of interstate wars since 1815 (Sarkees and
Wayman 2010), and that fraction contains some of the longest, bloodiest, and
most consequential conflicts of the modern era. When coalitions make demands
backed up by the threat of war, they sometimes win concessions peacefully;
other times, they must carry out their threats and wage war together. Further,
while most coalitions manage to keep their targets isolated, some provoke
balancing from third parties, expanding conflicts beyond their original participants. What accounts for these varied patterns of coalition-building, escalation
to war, and conflict expansion? And to what extent can a single theoretical
approach help shed light on them all? In this chapter, I contend that extant
work on military cooperation – chiefly focused on treaties of alliance and diplomatic multilateralism – is ill-equipped to answer these questions. I then outline
my alternative, as well as a new dataset, based on coalitions built in response
to specific crises, highlighting the role of decisions over military cooperation
and its maintenance in the course of individual conflicts, as opposed to prior

commitments and the sanction of international organizations, as is the case in the extant literature.

Formal alliances, in which states sign and ratify treaties that commit them to certain actions in the event of war (Morrow 2000), are the most commonly studied form of military cooperation. However, once general deterrence fails and crises arise, allied coalitions are actually less ubiquitous than those formed around no prior commitments. In the sample of late twentieth and early twenty-first century crises analyzed here, roughly three-quarters of coalitions involve *no* allied states. Allies may indeed side with one another in crises, though not necessarily in fulfillment of treaty commitments, and many coalitions that do involve allies include nonallied partners as well. For example, the Gulf War coalition of 1991 drew from regional powers such as Saudi Arabia, Syria, and Jordan, as well as NATO members – for example, the United States, United Kingdom, Turkey, and France – despite the fact that the conflict did not activate NATO's collective defense clause. Theories of alliances and alignment, which focus on the credibility of threats and the quality of cooperation (e.g., Benson 2012, Leeds 2003b, Smith 1995, Snyder 1997), are not designed to explain the multilateral politics of crisis-specific military cooperation – that is, the politics of military coalitions. I argue in this chapter that focusing on the unique dynamics of crisis-specific coalitions can offer better answers to the book's central questions about international cooperation, war onset, and conflict expansion than existing work on military cooperation does.

A second, more recent strand of literature examines the choice between unilateral and multilateral action, particularly for great powers like the United States that plausibly have the option to "go it alone" (e.g., Chapman 2011, Kreps 2011, Nye 2002, Thompson 2006, Voeten 2001). However, "multilateralism" in this context most often refers to military actions taken with the approval of international institutions like the United Nations Security Council (UNSC) and, often as a result, the support of certain other states. While UNSC support can under some conditions encourage states to join the same side in interstate crises (Chapman 2011, ch. 5), such support is often diplomatic as opposed to military in nature (Kreps 2011, ch. 4). As a result, some instances of military unilateralism, such as the American threat to invade Haiti in 1994 lest the governing junta relinquish power and restore the elected president, occur under the umbrella of diplomatic multilateralism; at the same time, other instances of diplomatic unilateralism, such as the Iraq War of 2003, are nonetheless useful – and illuminating – examples of military multilateralism. The distinction between diplomatic and military multilateralism is a critical one, because the latter figures more directly into states' short-run calculations of relative power and interest during international crises.[1]

---

[1] I do, however, make the case for accounting for the relationship between diplomatic and military multilateralism in empirical models of coalitional behavior, and Chapters 3–5 use an indicator of UN support as an explanatory variable.

In the remainder of this chapter, I outline the differences between military coalitions, treaties of alliance, and diplomatic multilateralism, making the case for studying the former as distinct from the latter two. Then, I use this discussion to establish a working definition of "coalition" as a particular form of military multilateralism. Next, I review relevant work on the dynamics of war and crisis bargaining in coalitional settings, identifying the first principles relevant for developing the theoretical models that I analyze in the following chapters. Specifically, two common factors – military power and the distribution of preferences within the coalition – make a substantial contribution to explaining the politics of (a) coalition bargaining, formation, and partner choice; (b) crisis bargaining, signaling, and the probability of war; and (c) the expansion of coalitional conflicts beyond their original participants. As discussed further later in the book, a theory of the expansion of conflicts also requires an integrated theory of coalitional durability, which I develop and test alongside predictions about conflict expansion.

I close this chapter by discussing the data collected for this project, detailing how the process of identifying coalitions and their relevant characteristics follows from the definitions and first principles defined here. I then use the data to establish a set of facts to be explained in subsequent chapters, showing not only that coalitions experience war and conflict expansion at different rates than single disputants, or "singletons," but also that these differences cannot be explained simply by differences in their observable shared characteristics, such as aggregate military power. In other words, coalitions are more than the sum of their parts; they do not simply behave like very powerful states. Rather, they require their own, more general sets of first principles, theoretical models, and explanatory variables, which I go on to develop in the following chapters.

2.1 DEFINING COALITIONS

A military coalition is a group of two or more states that makes a threat to use force together against another state (or states) in an international crisis (see also Wolford 2014a, 2014b). The threat can be explicit or implicit, deterrent or compellent, and the crisis may or may not ultimately escalate to war; the key element is the threat of collective military action should the coalition's demands be resisted, its threats ignored, or its warnings unheeded. Coalition members may contribute either military forces or territory for basing and staging support directly relevant to the threatened military action. At a minimum, this requires (a) the existence of a crisis or dispute in the shadow of military conflict and (b) multiple states taking the same side, not just diplomatically but militarily, before the crisis escalates to war. Every coalition has, at root, the same goal: "winning" (Riker 1962). However, before they can bring their pooled military resources to bear against the target, coalitions must also agree on the terms of their cooperation, which implies two types of conflict: disagreements between the coalition and its opponent, as well as disagreements between coalition members themselves over the distribution of the costs and benefits of joint action (cf. Snyder 1984, 1997).

By focusing on conflict-specific instances of military cooperation, my definition of military coalition satisfies Gamson's (1961) classic formulation of coalitions as "temporary, means oriented alliances among individuals or groups which differ in goals," whose members observe "tacit neutrality" on other, less immediate issues of concern (p. 374).[2] Likewise, Ward (1982, pp. 9–14) defines coalitions as issue-specific, time-limited, and exclusive in their membership, as opposed to alliances, which can (and often do) cut across sides in a given dispute (cf. Zinnes 1967) and may even be formed to restrain partners from engaging in disputes with one another (Gibler and Wolford 2006, Pressman 2008, Weitsman 2004). Though it shares Gamson's (1961) and Ward's (1982) focus on issue-specific cooperation, my definition allows me to problematize both the "tacit neutrality" at the heart of crisis-specific coalitions and the decision to cooperate with some state (or states) to the exclusion of others, shedding light on the attendant consequences for patterns of international war and peace.

In practice, coalitions take on a variety of forms. For example, a large coalition led by the United States threatened Iraq with war to coerce a withdrawal from Kuwait after its August 1990 invasion, while a smaller one – the United States, the United Kingdom, and Kuwait – threatened another war in 2003 over Iraq's compliance with UN-imposed limits on its development of weapons of mass destruction.[3] Another large coalition, made up almost exclusively of great powers, threatened military reprisals against China to accompany demands that it deal with the threat posed to foreign interests by the Boxer Rebellion in 1900. In their demands that Serbian forces withdraw from Kosovo under pain of air strikes and, later, a ground invasion, NATO countries constituted a coalition in 1998–1999, though it is notable that action was not taken as a result of the invocation of the alliance's central collective defense clause. Finally, the several joint French-Chadian threats of war over Libyan support of insurgents in the 1970s and 1980s also qualify as coalitions by my definition. On the other hand, the American threat to invade Haiti to restore ousted President Jean Bertrand Aristide, despite diplomatic support and the sanction of international organizations, remains unilateral under my definition, since the threatened invasion would have involved only American military forces, operating without foreign staging or basing support.

---

[2] Note that Gamson uses "alliance" differently than do most international relations scholars, focusing more on the short-run concern of active cooperation as opposed to the terms by which states commit to future actions.

[3] Kuwait satisfies the definition of a member of a military coalition because it provided critical staging and basing support. Other states that were diplomatically supportive but contributed neither military forces nor territory for basing and staging, however, do not satisfy the definition.

These examples also show that coalition members may contribute military forces directly, as the United Kingdom did during the 1991 Gulf War, as well as basing or staging rights, as did Turkey in the same war (see Chapter 3); each type of contribution has a material effect on the coalition's prospects for winning a military conflict, for which contributing states expect to receive some form of compensation, whether distributive or policy-related, for their efforts (cf. Riker 1962). It is worth noting that strictly financial support – as Japan provided during the Persian Gulf War of 1991 (Purrington 1992) – can also influence the chances of military victory; however, providing financial assistance does not carry with it the cost or the risk of active military participation in the war that, say, contributing troops or granting basing rights entails. Therefore, I do not consider financial support in the present analysis.[4]

In most international relations research, "coalition" is used interchangeably with "alliance," especially in the context of wartime cooperation or countering rising threats to the balance of power (see, e.g., Resnick 2010/2011, Snyder 1991, Walt 1987).[5] However, the two are analytically and empirically distinct (see also Ward 1982). Defensive and offensive alliances are *treaties*, formal promises to cooperate – that is, to form coalitions – in the event of war (Gibler and Sarkees 2004, Leeds et al. 2002, Morrow 2000), but whether the promised coalitions form depends on the activation of their treaties and fulfillment of their terms.[6] Contrasting the frequency of alliances in peacetime to the relative rarity of coalitions during 19th Century Europe, Gulick (1967, p. 78) notes that

...a listing of alliances would sound like someone reading from a telephone book, so bewilderingly abundant are they in the same period.

Though his definition of coalition differs from mine, Gulick's point about the frequency of alliances relative to their use in conflicts is illuminating. Many alliances, in particular successful defense pacts, are never invoked (cf. Smith 1995), and while some coalitions *do* form to honor formal alliance commitments, a large majority include nonallied states (Wolford 2014a, p. 147). Even some coalitions made up of allies fight together out of interests that fall outside the scope of their treaty commitments, such as the American-led wars against Iraq and the NATO interventions in Kosovo (1999) and Libya (2011).

---

[4] It is, however, an interesting avenue for further research, because financial support can help offset some of the costs of the war effort.

[5] Smith (1995), however, is an exception, noting that an alliance "contrasts with a coalition, which is a set of nations that simply fight together in a war, whether or not they had a previous agreement to do so" (p. 410).

[6] Neutrality pacts, on the other hand, can be understood as promises *not* to form certain coalitions under stipulated circumstances. Consultation agreements are more difficult to understand in the current framework, though the fact that they fall under the heading of alliances and not coalitions shows that my concept of coalitions is autonomous – coalitions are not merely a subset of alliances.

Therefore, while defensive and offensive alliances are peacetime promises to form coalitions that may or may not be fulfilled (Leeds, Long, and Mitchell 2000), coalitions are comprised of those states that *do* take the same side in a crisis, whether or not a prior commitment obligates them to do so. In starker terms, treaties of alliance are neither necessary nor sufficient for the formation of military coalitions. Italy's alignment decision in the First World War is instructive; it joined a coalition to which it was not committed by treaty, the Entente, in a fight against states to which it *was* formally committed in the Triple Alliance (Keegan 2000, Philpott 2014, Stevenson 2004).[7] As Snyder (1991) puts it, "[a]n alliance is a 'promise'," distinguishable from "the physical act of assisting others, whether or not in fulfillment of a prior promise." My definition of coalition is designed to facilitate analysis of the latter – "the physical act of assisting others" – not the former, though allied states can and do appear as coalitions in specific crises.

Other definitions of "coalition" exist, although most are designed to answer questions about rather different forms of military or diplomatic cooperation than those that are the focus of this book. Gulick (1967, pp. 77–89) identifies coalitions as groups of four or more states sharing the aim of preserving the balance of power among the great powers. In a pioneering analysis of the distribution of gains and losses after war, Starr (1972) identifies coalitions as states fighting on the same side in war, as does Smith (1995). In Chapter 5, I use a similar defintion, based on the Correlates of War data (Sarkees and Wayman 2010), to model the durability of war-winning coalitions. Morey (2011) also examines war coalitions, but his definition is narrower, requiring that states on the same side also exhibit a sufficient level of military coordination, such as joint planning or integrated command.[8] In contrast to both of these definitions, I focus primarily on coalitions at the stage of crisis bargaining, where their members take the actions that determine whether disputes ultimately escalate to war.

The distinction between prewar and wartime coalitions is far from trivial. Byman and Waxman (2002, ch. 6), for example, examine the problems faced by coalitions applying pressure to their targets once bargaining has already broken down in war – that is, once their initial coercive threats have failed and must be carried out. This, of course, is problematic for making inferences about whether building a coalition increases the chances of war over acting as a singleton because it begs the questions of why coalitions form in the first place, especially those that will exhibit major cooperative problems, and why their demands are resisted. If coalitional threats are heeded only when coalitions

---

[7] See Chapter 6 for a discussion of how Italy chose a relatively credible promise of compensation from the Allies over an incredible one from the Central Powers (see also Philpott 2014, pp. 79–80).

[8] Auerswald and Saideman (2014) study the problems of coalitional war in the context of alliance politics in the NATO operation in Afghanistan.

will be successful in war, then those coalitions we do observe fighting together will be a nonrandom sample of all possible coalitions; as a result, inferences about their military effectiveness will be difficult, if not impossible, to draw. Further, the target's decision to acquiesce is not the only selection process at play; states sometimes take great pains to choose their partners wisely, which means that the *least* militarily effective coalitions will never form in the first place. As such, this book's explicit focus on crisis-level bargaining and coalition formation holds out more promise of understanding the behavior of those military coalitions that do make it into the historical record.[9]

Other studies cast wider empirical nets, but none captures the crisis-specificity or exclusively military cooperation that animate my definition. Tago (2007) analyzes empirical models of broadly cooperative international endeavors, from military operations to humanitarian intervention, while Chapman (2011) operationalizes coalitions as made up of states taking the same side in crises, whether militarily or diplomatically, including those that join after bargaining breaks down in war. Finally, Kreps (2011) analyzes several American-led "coalitions of convenience," in which powerful states collect diplomatic or military partners for foreign "interventions," which by her definition do not include conflicts in which the powerful state fights over its own or neighboring territory (ch. 2). My definition, however, does not restrict the issues over which states contend, focusing on a wider variety of disputes that may escalate to war; it does, though, focus more strictly on military rather than diplomatic multilateralism. It also shares Kreps's concern, similar to Riker's (1962), over the trade-offs of military benefits and policy concessions entailed in coalition-building. For her, states that build coalitions trade time spent negotiating the terms of cooperation in return for military assistance – time that could otherwise be used confronting the enemy unilaterally. The implicit assumption of this approach is that, given sufficient time, any coalition is possible. However, coalition-builders vary in their ability to act as singletons (the United States is unique among even the great powers in this regard), and partners come at different prices, not only in terms of foregone time but also political goods – for example, shares of the spoils, of burdens, or of risk – that can have a significant impact on the dynamics and outcome of a crisis, as well as the feasibility of taking on any particular partner. In other words, the trade-off at the heart of coalition-building involves not only time spent waiting, which affects only the coalition-builder directly, but also the political values traded in return for military cooperation, which have a bearing on both the terms on which a partner is willing to cooperate and on which a lead state is willing to build a military coalition.

---

[9] Chapter 4, for example, analyzes the escalation of crises to war by deriving predictions from a theoretical model that includes both partner choice and a strategic target state, leading to empirical results that are at times surprising.

2.1.1 Coalitions versus Treaties of Alliance

Alliances have traditionally dominated scholarship on military cooperation, and one reason is the relative ease with which treaties can be observed and their characteristics measured (see Benson 2012, Gibler and Sarkees 2004, Leeds et al. 2002). This is especially true when compared to coalitions, which routinely have a less concrete empirical referent (Snyder 1991, p. 132). Another is the prominence of NATO and the American defense commitment to Western Europe after the Second World War, which prompted inquiry into the sources of credible deterrent commitments as the United States embarked consciously on its first extended foray into great power politics.[10] As such, while some work explores the question of which states are most likely to sign treaties of alliance with one another (Gibler and Rider 2004, Gibler and Wolford 2006, Kimball 2010, Lai and Reiter 2000, Poast 2012, Siverson and Emmons 1991), Cold War problems of extended deterrence – that is, how to deter threats to far-flung allies on another contintent – and long-term alliance management placed a premium on understanding the credibility of commitments to fight (Huth and Russett 1984, Morrow 1994, Snyder 1997), problems of abandonment and entrapment (Benson 2012, Snyder 1997), alliance reliability (Gartzke and Gleditsch 2004, Leeds 2003a, Leeds, Long, and Mitchell 2000), burden-sharing (Bennett, Lepgold, and Unger 1994, Olson and Zeckhauser 1966), and alliance duration (Bennett 1997, Gaubatz 1996, Leeds and Savun 2007, Morrow 1991).

Given its often heavy focus on deterring challenges to the status quo, the key explanatory variables in this body of scholarship are the peacetime costs of formal commitments (Morrow 1994), the reputational consequences of reneging on those commitments (Gibler 2008), domestic institutions (Gibler and Wolford 2006, Leeds, Mattes, and Vogel 2009), and the foreign policy goods exchanged by allies (Leeds, Long, and Mitchell 2000, Leeds and Savun 2007, Morrow 1991, Poast 2012).[11] However, alliances formalize and commit states to promises of *future* behavior. As a result, the theories designed to explain them do not apply well once (a) commitments are invoked by an event satisfying the *casus foederis*, setting the stage for negotiating the specifics of cooperation – which even long-committed allies find to be challenging – or (b) states form coalitions with nonallied partners, for whom even more elements of the cooperative military relationship remain to be negotiated.

---

[10] Indeed, even after the Cold War, debate continued over the future role of NATO after some scholars predicted its imminent demise (see, e.g., Mearsheimer 1995, Wallander 2000).

[11] That Gibler (2008) finds evidence that leaders violating treaties are less likely to form subsequent alliances is remarkable, because the data-generating process stacks the empirical model against finding such a pattern; if leaders strategically avoid the most costly violations, then finding evidence of reputational costs should be extremely difficult (cf. Schultz 2001). That this relationship was uncovered, then, suggests that the costs of violating alliance treaties are even larger than they appear in the observational record.

Rather, as states choose sides during a crisis in the shadow of possible war, a unique and altogether different set of questions becomes relevant, including (a) how coalitions are formed, (b) how much (and which) partners are to be compensated for their cooperation, and (c) how such choices affect crisis escalation and expansion. To be sure, existing alliance commitments are likely to play a role in the calculus, because allies share common enemies, may have experience fighting together, and may pay substantial reputational costs for failing to cooperate with one another (for a longer discussion of such differences, see Chapter 6). Nonetheless, alliance commitments are only one of a wide array of factors that shape the process of military multilateralism. Therefore, I argue that a useful theory of military coalitions must proceed from a different set of first principles than those with which the alliance literature is concerned, particularly because allies face the same, heretofore underexplored, set of questions that nonallied states face in negotiating, securing, and maintaining military cooperation when faced with a crisis – differing only in degree and not in kind.

First, where treaties of alliance are of necessity incomplete contracts, stipulating behaviors for an uncertain future (Benson 2012) and trading off broad elements of security and autonomy (Morrow 1991), coalition formation requires more immediate coordination over mutually acceptable aims, escalation levels, bargaining strategies, and shares of the spoils of victory (see Starr 1972).[12] In other words, coalition negotiations, whether or not they occur between allies, involve trades over military capabilities and policies *specific to the crisis at hand*, and those specifics are generally difficult to anticipate.

Coalition negotiations are often over the very issues that incomplete alliance contracts exclude from their terms. As mentioned earlier, long after binding themselves by treaty to the defense of Poland – and each other – "[i]t was agreed [in late 1939] that the British and French would share the cost of the war effort sixty-forty, proportions reflecting the relative size of their economies" in the Second World War (Hastings 2012, p. 27). Similarly, the Locarno Treaty in 1925 pledged France and Great Britain to the maintenance of Belgian territorial integrity, but the "logistical, intelligence and operational dimensions to a punitive action if Germany reneged on her signature were never worked out" (Alexander and Philpott 2002a, p. 2). Likewise, even after signing a military pact early in 1950 promising mutual assistance in the event of war, Communist China and the Soviet Union entered fraught negotiations over precisely how the latter would assist the former once it entered the Korean War; these negotiations resulted only in the stationing of Soviet aircraft in Manchuria in a

---

[12] Through a similar exercise of incomplete contracting, alliances may also be signed with an eye to pacifying or controlling their own members (Gibler and Wolford 2006, Johnson and Leeds 2011, Mattes and Vonnahme 2010, Pressman 2008, Weitsman 2004), though my focus here is on partner choice in terms of strategies for facing a specific target, for which the closest analogue is an externally oriented alliance.

limited, effectively non-belligerent role, in contrast to Stalin's initial promises of more extensive logistical and air support (Stueck 1995, pp. 38–39, 100–101). Austria-Hungary and Germany, likewise committed to fighting together, negotiated over the terms to be put in front of Serbia during the July Crisis of 1914 and the disposition of military forces in the event of war with Russia (Clark 2012, Fromkin 2004, Hastings 2013). Indeed, Benson (2012, p. 41) notes that treaties of alliance in which states lay out specific demands of their targets are exceedingly rare. However, these negotiations over aims and bargaining positions with which my theory is concerned are of primary importance once crises emerge, general deterrence fails, and states begin the process of aggregating military and diplomatic power; the specific issues of future crises are not easily anticipated in the design of prior commitments.

Second, given the alliance literature's focus on the credibility of promises and threats, it generally treats the probability of war through the prism of uncertain alliance reliability (e.g., Smith 1995). In coalitional crises, however, general deterrence has already failed, and coalitions still face the challenge of signaling their resolve over the issues of contention, which may pose difficulties that states acting alone do not face (Christensen 2011, Lake 2010/2011). Owing to divergent preferences and an uneven distribution of the costs of war, negotiations over dealing with common enemies are often difficult. For example, in the Berlin Crisis of 1961, NATO countries considered at length a range of options to signal their resolve over protecting the Cold War status quo (Aono 2010, Dallek 2003, Freedman 2000, Kempe 2011), prior obligations to support one another in the event of armed attack notwithstanding. Likewise, an endless series of conferences over common military strategy – few of them directly fruitful – characterized Allied warmaking throughout the First World War (Hastings 2013, Philpott 2009, 2014). Even if the threat to fight together is perfectly credible, allies must still negotiate the particulars of the war effort, and success or failure at this point can alter an opponent's beliefs about the relative desirability of honoring and refusing to comply with the coalition's demands.[13] Put differently, reliable alliances can be effective at bargaining and fighting together, or they can be ineffective, and which of these happens to be the case will depend in large part on how military cooperation is negotiated and tailored to the crisis at hand. Therefore, understanding bargaining and signaling in coalitional crises requires a theory of whether and how multiple states can signal resolve to fight over specific issues – a dynamic that has eluded consideration in most studies of military cooperation. To the extent that it has been explored, coalitions are generally considered, almost by construction, a hindrance to credible communication as opposed to a solution.

The principal explanatory factors in theories of alliance formation and extended deterrence thus provide limited help in understanding when and how

---

[13] Philpott (2014) provides a sprawling, insightful discussion of how knowledge of intracoalitional politics affected war-fighting decisions on both sides of the First World War.

states form coalitions in specific crises.[14] This imposes further limits on our understanding of the conditions under which crises expand to include other states or escalate to war, because they do not explain why and how states with different preferences – over war, peace, and continued cooperation – come to form coalitions, negotiate, and arrive at a collective military strategy. A strict focus on alliances, after all, might lead one to expect Italy to fight alongside its Triple Alliance partners in the First World War, but the specifics of the crisis, such as Italy's own territorial ambitions around the Adriatic, led to a rather different pattern of coalition formation. To better understand these processes, I turn later to a discussion of how two principal factors – the military power of, and the distribution of preferences within, the coalition – can shed a new and unique light on the process of military multilateralism.

2.1.2 Military versus Diplomatic Multilateralism

We can also distinguish military coalitions from what might best be termed diplomatic multilateralism, which involves securing the approval of international institutions like the United Nations for military action that may, but need not, facilitate the process of building military coalitions (Chapman 2011, Kreps 2011, Tago 2007, Voeten 2001).[15] Diplomatic multilateralism, then, entails political support channeled through formal institutions, not merely verbal statements of support from sympathetic countries – such as a number of states that acceded to joining the United States' "coalition of the willing" over Iraq's WMD programs in 2003 (see Leetaru and Althaus 2009) or the large number of countries that, with the outcome fairly visible, declared war on Germany in the waning months and years of each of the world wars.[16]

Just as Cold War concerns over the credibility of extended deterrence highlighted the value of studying alliances, the beginning of the United States' "unipolar moment" (Krauthammer 1990) and the end of bipolarity – especially its ever-present threat of offsetting superpower vetoes in the United Nations Security Council – appeared to open a new era in which international institutions would play an important role in the authorization and legitimation of the use of force (Chapman 2011, Thompson 2006, Voeten 2005). Adding relevance to this line of inquiry was the highly visible public debate over the United States' launching of the 2003 Iraq War without the elusive "second resolution" from the Security Council that would have added an additional layer of approval for the invasion (see Keegan 2005, Nye 2002). Nonetheless, while

---

[14] However, the existence of a specific, activated commitment may play a role in subsequent war-joining (Leeds 2003a, Reiter and Stam 2002).

[15] See also Wolford (2013, p. 296) and Ikeda and Tago (2014) for the use of "diplomatic multilateralism" to describe this phenomenon. The latter also define "operational" multilateralism in a way similar to my concept of "military" multilateralism.

[16] A decision that, interestingly, would bear on debates at the Yalta conference over how to apportion influence in the postwar world (see Plokhy 2010).

the approval of certain types of institutions can encourage states to side with others in international disputes (Chapman 2011, ch. 5), my focus is on strictly military cooperation, allowing me to bring into focus some events that are often labeled "unilateral" in the diplomatic sense yet still implicate questions of the negotiation and maintenance of military cooperation.

Indeed, "multilateral" itself connotes only the involvement of multiple states on one side in a given dispute.[17] However, it is often used as shorthand for diplomatic multilateralism, which scholars have linked with phenomena including both joint military threats and diplomatic support under the aegis of a formal international institution (Chapman 2011, Kreps 2011). Military coalitions as defined here may or may not have the legitimating support of international institutions; the Axis, after all, was multilateral, but its withdrawal from the League of Nations rendered moot any possibility of diplomatic multilateralism. My definition only requires the threat of cooperative military action against a common target or targets in international crises. Two examples are instructive. In both the Kosovo War of 1999 and the Iraq War of 2003, American-led coalitions used military force without the formal approval of the UNSC due to credible veto threats from Russia and France, respectively. While the Kosovo War can be considered diplomatically multilateral by virtue of the involvement of a formal institution (NATO), the Iraq War is very often labeled a "unilateral" use of force, despite the fact that the United States received meaningful military cooperation from both Kuwait and the United Kingdom.[18] While the Iraq War certainly fails some tests of *diplomatic* multilateralism, it does not fail the test of *military* multilateralism, which is the central feature of coalitions as I define them here. This distinction highlights what is at stake, from an analytical perspective, in how scholars choose to define and operationalize uni- and multilateralism. Indeed, as I discuss in Chapter 3, understanding the failed negotiations over Turkey's potential membership in the coalition for the 2003 Iraq War requires a focus on military – not diplomatic – multilateralism.

The wars over Kosovo and Iraq both include what are, for my purposes, military coalitions. In this context, the role of international institutions is perhaps best viewed as a potentially intervening variable in the process of coalition formation. Coalitions are by definition multilateral in the military sense, but they need not be diplomatically multilateral. Further, since coalitions are defined by threats of joint military action, some events, notably the American threat to invade Haiti in 1994, which Kreps (2011, ch. 5) identifies as multilateral, do not properly involve military coalitions. To help fix ideas, Table 2.1 provides a

---

[17] But see Keohane (1990), who proposes a three-state minimum for defining multilateralism as against bilateralism, and Corbetta and Dixon (2004), who follow that standard in a model of multilateral dispute participation.

[18] Given the loaded nature of "unilateral" as a descriptor of foreign policy, I instead use the term "singleton" to denote a state that faces its target alone, which more easily reflects the difference between states that do and do not build coalitions. I am indebted to an anonymous reviewer for suggesting this term.

TABLE 2.1. *A Typology of Multilateralism*

| | Military | |
Diplomatic	No	Yes
No	Invasion of Grenada	Iraq War
Yes	Haitian Intervention	Persian Gulf War

simple breakdown of how these definitions apply to four separate conflicts: (a) the 1983 American invasion of Grenada, a conflict that was neither diplomatically nor militarily multilateral; (b) the 2003 Iraq War, which was multilateral only in the military sense, as it lacked formal approval of an international institution; (c) the United States' Haitian Intervention of 1994, which was unilateral militarily but received support from the United Nations; and (d) the Persian Gulf War of 1991, which satisfies both definitions of multilateralism, with multiple countries making military contributions and explicit authorization to use force from the United Nations. Therefore, while I account for the role of diplomatic multilateralism in the empirical models of Chapters 3–5, my primary focus is military multilateralism, which can – and, as shown later in the book, often does – proceed without its diplomatic counterpart.

2.2 THE PROCESS OF MILITARY MULTILATERALISM

Military multilateralism is characterized by three distinct but interrelated stages that begin as soon as a crisis is triggered between a set of primary disputants: (1) coalition formation, (2) crisis bargaining, and (3) conflict expansion. In the first stage, a state in a crisis or dispute chooses whether or not to form a coalition, which requires bargaining with one or more potential partners over the terms of military cooperation. In the second stage, the coalition turns to bargaining with its target, adding a dyadic element that creates strategic linkages between the distribution of preferences within the coalition and the international distribution of military power. At this stage, the need to maintain cooperation shapes the coalition's strategy for bargaining with its target, in particular the means by which it signals its willingness to fight if its demands are not satisfied. In the third stage, the conflict may expand to include additional states through balancing and the formation of counter-coalitions, undesirable from the coalition's perspective, in a process that can also be linked to coalitional power and preferences; specifically, outsiders concerned that coalition members will sustain military cooperation into future crises sometimes intervene against those that will be durable enough to pose a future threat. Figure 2.1 characterizes the temporal sequence of events and the chapters that cover them, from stages 1 through 3 to the post-conflict question of coalitional durability, which – since it is related to stage 3 indirectly – I also take up in Chapter 5.

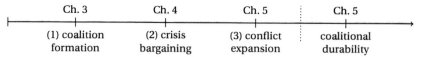

FIGURE 2.1. The stages of military multilateralism

There exist small and loosely connected literatures dedicated to the first two stages of military multilateralism, and I review them in this section at some length, highlighting the central roles of (a) the distribution of preferences within coalitions and (b) coalitional military power, both aggregate and relative, in understanding outcomes at each stage of the process. These two concepts provide anchors to existing work on, respectively, the politics of cooperation within coalitions and international crisis bargaining with their opponents. Then, I show how their interaction might explain a range of behavior in crises for which neither factor accounts on its own, including the final phase of conflict expansion and, after victory, the durability of cooperation over a postwar peace.

2.2.1 Coalition Formation

Interstate crises and disputes present their participants with a series of questions: How much military assistance do they want, what are they willing to pay in return for it, and who, if anyone, will cooperate at an acceptable price? States that would build coalitions thus face a basic trade-off in choosing those partners with whom they wish to cooperate (if any) and ensuring their cooperation in the crisis and possible war. Short of a perfect alignment of preferences and incentives, this may not be easy: states differ in their evaluation of the status quo, valuation of the stakes of the crisis, willingness to bear costs and tolerate risks, and preferred strategies for bargaining with their targets. Just as they do when building alliances (see Morrow 1991), states consider both potential military capabilities and specific contributions such as basing or staging rights when identifying potential partners, but ensuring their cooperation in costly endeavors such as crisis bargaining and war may require side payments and/or policy concessions that coalition-builders would, all else being equal, prefer not to make. Any coalition's goal is "winning" (cf. Riker 1962), or producing a favorable political outcome, but sometimes improved chances of victory come at an unacceptable price.

The price of compensation aside, the potential benefits of cooperating with coalition partners are considerable. First, coalitions can increase a state's available military power (Morgenthau 1967, Walt 1987, Waltz 1979), which may derive from the strict aggregation of capabilities as well as contributions such as basing or staging rights, both of which can act as force multipliers without directly adding to material capabilities. In the Iraq War of 2003, for example, the United States and the United Kingdom provided combat power, while

Kuwait contributed the overland route for the invasion. Second, coalitions also lower the costs of coercion and enhance capabilities through burden-sharing and specialization (see Kimball 2010, Lake 1999, Morrow 1993). The Allied coalition in the opening year of the First World War, for example, relied chiefly on French and Russian land power and on British naval power (Hastings 2013, Philpott 2014, Stevenson 2004), increasing the total combat power of the coalition at lower cost than would be possible absent specialization.[19] To secure these benefits, however, a potential coalition-builder must secure the cooperation of partners that may not share its interests or its resolve, requiring compensation in the form of side payments or policy concessions over the management of the crisis.[20]

Compensation comes in a variety of forms, from compromises over war aims and shares in the fruits of victory to future cooperation and promises of loans, indemnities, or other financial transfers (see Riker 1962, Starr 1972).[21] Several twentieth-century conflicts provide examples of each. First, the United States carefully calibrated its war aims, agreeing not to topple Iraq's government in 1991, lest it forego critical Saudi staging support for Operation Desert Storm; ultimately, the United States limited its goals to the restoration of Kuwaiti sovereignty and the degradation of Iraqi military capabilities (Bush and Scowcroft 1998, pp. 313, 491).[22] Next, during interventions in the Yugoslav civil wars, the United States agreed to limits on the air campaigns over Bosnia in 1995 (Papayoanou 1997) and Kosovo in 1999, even ruling out the early threat of a ground campaign in the latter (Clark 2001, chs. 6, 7), in order to ensure the cooperation of NATO partners in both operations. Likewise, American decision makers agreed to limit escalation in the Berlin Crisis of 1961–1962, opting against sending ground forces across the intra-German border (Aono 2010, pp. 326, 334), striking a delicate balance between convincing its NATO coalition partners that the United States "would not be rash," even as "the Soviet Union had to be persuaded that it just might be" (Freedman 2000, p. 93).[23]

A similar story can be told about Italy's entry into the First World War on the side of the Allies, despite a prior alliance commitment to the Central

---

[19] Eventually, though, Britain would raise and deploy a sizable army to the Continent.

[20] Diplomatic multilateralism can, of course, provide military benefits by reassuring other states of benign intent or provide political cover, lowering the costs of war by loosening domestic constraints or discouraging potential balancers from joining the war on the side of the target (Chapman 2011, Thompson 2006, Voeten 2005). I explore this possibility explicitly in later chapters. However, see Chapman and Wolford (2010) on the consequences of multilateral approval for crisis bargaining postures and the probability of war.

[21] Riker (1962, pp. 115–120) identifies three ideal types of side payment – "contingent payments out of profits," "payments out of working capital," and "payments out of fixed assets" – and those discussed here are closest in nature to the first two.

[22] See also Atkinson (1993, pp. 298, 299) and Kreps (2011, p. 27).

[23] See Chapter 4 for a lengthier discussion of the crises over Berlin and Kosovo.

Powers, which was secured by promises in the secret Treaty of London that it would receive "most of the Austrian territory it wanted, together with the Dodecanese islands in the eastern Mediterranean" (Keegan 2000, p. 226).[24] Whether Italy would receive all of its promised share of the spoils of victory after the fact – it did not – is a different question. Finally, in return for Soviet entry into the war against Japan in the Pacific theater and territory for strategic bomber bases in Siberia following Germany's defeat in the Second World War, the United States promised to fulfill "a huge tabulation of supply demands" and to recognize Soviet influence and territorial interests in the area lost after the Russian defeat at the hands of a rising Imperial Japan in 1905 (Frank 2001, p. 31–32).[25]

In many cases, on the other hand, would-be coalition-builders may decide that some contributions are not worth the necessary concessions. In 1951, the United States viewed a Pakistani military contribution to the Korean War coalition as highly attractive from a diplomatic standpoint, yet the likely military contribution proved prohibitively small relative to Pakistan's requested compensation: (a) an American defense commitment and (b) support against India in the disputed region of Kashmir (Stueck 1995, pp. 72–73). Many countries, in fact, hoped to reap rewards for participation in the United Nations forces during the war, particularly in the form of military aid and defense pacts with the United States, which accepted proffered support carefully, with an eye to minimizing the costs of compensating potential partners (Stueck 1995, ch. 2). Half a century after Korea, in securing limited staging and overflight rights from Pakistan for the invasion of Afghanistan in October 2001, the United States agreed to a package of financial assistance and to limits on overflight and staging privileges – although it refused most of Pakistan's demands aimed at undermining India in the region (Jones 2010, pp. 88–89). Two years later, the United States turned down the chance to open a northern front in the Iraq War, which could have forced Iraq to divide its forces and risk a quickened collapse, rather than promise to Turkey the share of the postwar spoils of victory it would have required in return: substantial financial compensation and the right to enter and establish order in Iraqi Kurdistan (Gordon and Trainor 2006, pp. 42, 115). Instead, the United States offered significantly less, which proved insufficient to overcome Turkish parliamentary opposition to joining the coalition (Keegan 2005, Migdalovitz 2003).[26]

---

[24] See also Stevenson (2004, pp. 90–92).

[25] While the Soviets did enter the war against Japan soon after Germany's defeat, many of the terms Stalin demanded, including a resumption of rights at Dairen and Port Arthur, failed to materialize, partially because those involving Chinese assent required the acceptance of Chiang Kai-Shek, whose Nationalist forces were in the final stages of the civil war against the Communists (see Frank 2001, Paine 2012).

[26] See Chapter 3 for a more detailed discussion of these events in comparison to the successful negotiation of terms of Turkish cooperation in the 1991 Persian Gulf War.

Thus, coalition-building requires compromise and balance among diverse interests, because states must be compensated for divergent preferences in order to make costly contributions to the coalition's military effort. They might value the issues at stake differently, vary in their sensitivity to the destructiveness or risks of war, or face different domestic constraints in the form of public opinion or budgetary limits. This can require policy compromises of a would-be coalition-builder, who weighs these sacrifices against the increased coercive power – that is, an increased chance of winning – it might enjoy as the result of cooperating with coalition partners. In other words, coalitional bargaining confronts a state with the option of bolstering its bargaining position against the target at the cost of compensating partners, or foregoing that improved position in order to preserve control over political aims and the spoils of victory. How states manage this trade-off in the context of specific crises is not well understood; rather, extant research typically takes partner choice as given – as well as the rejection of the coalition's initial demands – and analyzes the problems of either collective action or divergent preferences from that point forward (e.g., Auerswald and Saideman 2014, Byman and Waxman 2002, Choi 2012). However, this begs the critical questions of (a) why coalitions with divergent preferences form in the first place and (b) the conditions under which their demands are rejected, forcing them to carry out their threats of war.

By focusing on the interaction between divergent foreign policy preferences, which shape the costs of cooperation, and military power, which shapes its benefits, Chapter 3 provides an explanation of how diverse coalitions form in the first place. This sets the stage for explaining the process of coalitional crisis bargaining, particularly how outcomes depend on a negotiated balance of diverse interests, discussed in Chapter 4, as well as the expansion of conflicts to include counter-coalitions, driven by yet another interaction between power and preferences, in Chapter 5.

2.2.2 Crisis Bargaining

After coalitions form and agree on the terms of their cooperation, they turn to bargaining with their targets. In contrast to demands made by singletons, coalitional demands and threats reflect a negotiated balance among diverse national interests. However, most extant work remains unclear on just how this affects the dynamics of war and peace. States often disagree over the distribution of territory, rights, influence, or policy, and crisis bargaining is the search for peaceful settlements of these disputes in the shadow of war (Fearon 1995, Powell 1999), where outcomes are shaped by both (a) relative military military power and (b) beliefs about each side's resolve, or willingness to fight (Fearon 1994, Morrow 1989). Most work on crisis bargaining focuses on dyadic interactions (e.g., Fearon 1997, Schultz 1998, Slantchev 2005), but the diversity of preferences within coalitions suggests that the dynamics of signaling resolve – and even the effects of relative military power (cf. Croco and Teo 2005, Sobek and Clare 2013) – may be quite different across dyadic and multilateral contexts (Christensen 2011, Lake 2010/2011). As discussed in more detail later in this chapter, crises involving coalitions on at least one side are, on average, significantly more likely to escalate to war than are crises that do not involve coalitions.

Indeed, scholarship is often skeptical that coalitions can apply effective coercive pressure to their targets (e.g., Auerswald and Saideman 2014, Byman and Waxman 2002, Choi 2012, Tago 2009), which could well undermine the credibility of their initial threats and make their opponents more likely to resist their demands.[27] Many of these suspicions stem from an application of the basic logic of collective action, in which states may attempt to free-ride on the efforts of others, underproviding military contributions or simply reneging on commitments in order to husband their own resources or reduce their exposure to casualties and risk (Olson 1965, Olson and Zeckhauser 1966).[28] This, of course, is a problem that a coalition's opponents can anticipate, yet the majority of this work focuses on cooperative actions such as joint military operations, typically viewed as failures in absolute terms, that occur only *after* a coalition's demands have been rejected and their threats of war called in. This begs the critical question of why, if coalitions are so prone to collective action failures, states build them at all, which obscures the conditions under which coalitional crises either escalate to war or end in peaceful settlements, an area in which a different body of scholarship is equally skeptical.

Despite some minimum level of shared interests in a crisis – say, reinstating a conquered state's sovereignty, ending genocide or ethnic cleansing, replacing a target state's government, capturing new territory, or securing indemnities – coalitions nonetheless exhibit a significant amount of preference diversity. This is especially so when the costs and risks of war fall differently across their members. Partners disagree over what threats to make, what signals to send, and, in general, how to deal both diplomatically and militarily with their targets (see Auerswald and Saideman 2014); they may prefer ground wars to air campaigns, limited war aims to expansive ones, longer deadlines for compliance to shorter ones, and so forth. In such cases, some analysts suggest that the challenges of signaling resolve, already considerable in a dyadic interaction (Fearon 1995, 1997), are magnified, because states must consider the interests and potential reactions of either their partners or other international audiences (Christensen 2011, Russett 1963).

---

[27] For a similar argument applied to multilateral economic sanctions, see Miers and Morgan (2002). For evidence that coalitions tend to win a disproportionate number of the wars they are fight, see Morey (2011).

[28] It is important to note that collective action problems do not emerge from divergent preferences – in fact, divergent preferences can help solve them (Olson 1965, Sandler 1992, Sandler and Hartley 2001) – so I focus instead on problems more directly linked to preference diversity, in particular some partners' inability to commit to fighting if the war will be too costly for them.

l, "size" or "numbers" is considered a hindrance to solving informational problems in a variety of contexts. Lake (2010/2011), for example,
claims that coalitions face inherent signaling problems as a direct result of their
divergent preferences, a problem he argues was manifest before the Iraq War of
2003. Bellamy (2000) argues that a similar lack of consensus undermined the
credibility of NATO threats before the 1999 Kosovo War, and Clark's (2001)
account of intra-allied bargaining in the lead-up to war is flush with examples
of intramural disagreement over how, exactly, to threaten to attack Serbia.
Finally, structural accounts of international politics that focus on the number
of great powers argue that estimates of resolve or likely military outcomes may
become more complicated when more potential players are involved (Huth,
Bennett, and Gelpi 1992, Mearsheimer 2001, Waltz 1979), making it more
difficult to gauge what settlements one's opponents might accept in lieu of war.

Nonetheless, as shown in Chapter 4, unconditional claims about the link
between coalitions and information problems are accurate only under particular sets of conditions, which can only be uncovered by focusing on intra-
coalitional politics. The literature on deterrent alliances, for example, argues
that the presence of multiple actors can make for *more* credible threats to fight
than states acting alone, enhancing deterrence and reducing the chances of
war by resolving an underlying informational problem (Leeds 2003b, Morrow
1994, 2000).[29] In this case, more actors might actually reduce the problems
caused by uncertainty by introducing observable factors, such as military capabilities, that can swamp uncertainty over unobservable factors, such as resolve
(cf. Reed 2003). Russia, Germany, and France, for example, had little trouble convincing Japan to yield some of its gains from the Sino-Japanese War
during the Triple Intervention of 1895 (Iklé 1967). How, then, are we to rec-
oncile the claim that increasing the number of actors exacerbates information
problems with the more sanguine picture painted by alliance scholarship? Or
might both be contingently true, depending on features of particular crises and
intra-coalitional politics?

In models of crisis bargaining with private information, peaceful outcomes
often rest on a resolute state's ability to distinguish itself as such with a costly
signal, given that an irresolute state would wish to bluff about its resolve (see
Fearon 1997, Jervis 1970). A resolute disputant must pay some price that
an irresolute type is unwilling to pay, influencing an opponent's estimate of its
resolve such that the opponent offers acceptable terms. Otherwise, an opponent
may run a risk of war, offering harsh terms if it is optimistic enough that the
state in question truly is irresolute. Yet signals sent in crises are relevant not
only to one's opponent but also to third parties, who can in some cases stand
in the way of effective signaling. Fearon (1997) conjectures that the failure of
some signals to convey resolve is less important "than the effect achieved on

---

[29] But see Vasquez (2009, ch. 5) for the claim that alliances do not deter conflict but merely deepen
mutual suspicion and hostility – at least between equals – which then lead to "wars of rivalry."

various other audiences" (p. 84), while Russett (1963) contends that "pressures from the 'attacker's' own allies" (p. 102, fn. 8) affected escalatory behavior in Cold War–era crises over the Bay of Pigs and the Suez Canal.[30] As discussed further in Chapter 4, the United States faced just such a dilemma during the Berlin Crisis of 1961–1962; concerned with the reaction of allies nervous about the costs of a potential war, it limited escalation levels in aid of preserving military cooperation, despite the risk of failing to demonstrate resolve (Aono 2010, Freedman 2000, Kempe 2011). The question, then, is how this internal haggling over external bargaining strategies shapes the subsequent course of the crisis.

Compared to states acting alone, coalitions face a unique set of challenges that stem directly from a coalition-builder's trade-off between aggregating military power and ensuring the cooperation of partners with divergent preferences. Taking on partners with diverse preferences implies disagreement over political aims, escalation levels, and/or the division of the spoils of victory – disagreement that can undermine the ability to send credible signals of resolve when cooperation necessitates compromises or concessions over coalitional strategy. Coalition leaders, for example, might prefer to make costlier threats than their partners, whose threat to deny military cooperation gives them leverage to restrain the lead state and, as a result, "water down" the coalition's threats or undermine its signaling strategy. However, the extant literature takes signaling failures as given, begging the linked questions of (a) why coalitions form around weak threats and (b) why their demands are rejected in the first place. By analyzing problems of coalition formation and crisis bargaining in an integrated theoretical model, Chapter 4 shows that the involvement of coalition partners can either facilitate *or* undermine credible signaling, depending on features of the distribution of power.

2.2.3 Conflict Expansion

While power and preferences play immediate roles in both coalition formation and crisis bargaining, they also factor into the final stage of military multilateralism, where the conflict expands to include new belligerents. When third-party states choose whether or not to balance against possible future threats, they set the stage for localized conflicts to become wider regional or global conflagrations, and an explanation of why coalitions, not just singletons, fail to keep their targets isolated can make a unique contribution to our understanding of war expansion. In 1914, for example, Germany and Austria-Hungary quite famously failed to keep the latter's war on Serbia localized (Hastings 2013,

---

[30] Fearon's (1997) conjecture involves signaling private information to multiple audiences, a possibility I do not explore in the present model. Rather, in the model of Chapter 4, a third party is aware of the sender's type but has preferences over military consequences of the signal sent.

ch. 2), to disastrous effect. As noted in Chapter 1, the most destructive multilateral wars feature coalitions on both sides – the world wars are obvious examples – yet much of the literature on balancing, alignment, and conflict expansion is rooted at the level of the singleton, *not* the coalitions that represent the initial step toward wider conflict expansion; I show in Chapter 5 that this omission is not without consequence.

Before states balance against a potential future opponent by siding against it in a conflict today, they try to assess both the opponent's power and the extent to which its intent is threatening. While this is true for both singletons and coalitions as the possible targets of balancing, the latter represent unique concentrations of power that pose greater post-conflict threats to outsiders than their members would individually. This may be true because a coalition leader retains capable partners or valuable basing/staging grounds, or because victory in war changes the facts on the ground, placing coalition members in a new, more powerful military position; fear of precisely this eventuality was what drove British fears about the consequences of a Central Powers victory over the Entente in 1914. Therefore, the final stage of military multilateralism is the potential expansion of the conflict, where third parties may join the target's side in order to balance against a coalition's looming future threat. Virtually all extant treatments of balancing and alignment focus on the threats posed by single states (see, inter alia, Christensen and Snyder 1990, Mearsheimer 2001, Powell 1999, Walt 1987, Waltz 1979), but the unique features of coalitions – not only their military power but also the diversity of their preferences – suggest that the process by which threatened third parties make alignment decisions with respect to coalitions may be different from the same process with respect to singletons (cf. Wolford 2014a).

When making calculations over balancing or remaining neutral in a conflict between other states, a fearful third party

faces a difficult inference problem when it sees one state attack another. Does the attack indicate that the attacker is generally aggressive and willing to use force or just that it is dissatisfied with the particular state it attacked? Does the attacker have limited or unlimited ambitions? (Powell 1999, pp. 193, 194)

In other words, will the attacker stop once it has secured victory, or will its appetite grow with the eating? Cheap-talk claims of benign intent in this context, where truly aggressive states have strong incentives to lie about their willingness to observe restraint, are unlikely to reassure mistrustful third parties (see Kydd 2005). Absent some credible reassurance that passivity today will not be rewarded with victimization in the future, states may join the conflict on the side of the coalition's target. This, in turn, has led scholars to examine the conditions under which states can use international institutions to signal their preferences or alter their own incentives for the exercise of power. A fearful observer may draw on an attacker's reputation and institutional commitments

(Ikenberry 2001), as well as the statements of consultative international organizations (Voeten 2005), when assessing its future willingness to use force or the expansiveness of its aims. While these same indicators are surely also useful for judging a coalition's future intentions, as well as predictors of coalition formation (Chapman 2011, ch. 5), some unique features of military coalitions suggest that the observer's inference problem may be more complicated in a multilateral context.

The reason is simple: to pose a future threat to third parties, coalition members must not only have similar interests over coercing additional targets but also maintain cooperation beyond the current crisis, avoiding conflict over the spoils of victory and acting as a coalition again in the future. Individual states face no such problems; either they are willing to seek concessions from a target in the future, or they are not. When coalitions are a potential threat, however, the issue of sustaining military cooperation emerges, and fearful third parties will be especially concerned over whether a coalition will stand or divide against itself, since intervening in another state's conflict in the present is wasteful if the coalitional threat will end up dissipating on its own in the future. If third parties find coalition members more threatening together than apart, then assessments of coalitional durability have consequences for future bargaining outcomes, which figure prominently in the logic of alignment decisions (see Powell 1999, ch. 5). While aggregate power affects the size of future threats, the distribution of preferences within the coalition affects its choices over staying together – that is, the probability with which a future threat emerges.

Thus, in Chapter 5, I present a model that combines three elements: (a) coalition formation, (b) a third-party observer's choice of balancing against a coalition or remaining neutral, and (c) a coalition's choice of disbanding or staying together. As with the first two models, the answer turns on both the coalition's military power and the distribution of preferences within it. Militarily powerful coalitions are uniquely capable of discouraging balancing when they have diverse preferences because they are likely to disband on their own in the future, yet weaker coalitions are actually *more* likely to provoke balancing when they have diverse preferences. By focusing on this crucial stage of conflict expansion, when a coalition's opponent may or may not receive assistance, Chapter 5 identifies a substantial – and potentially consequential – gap in our understanding of how conflicts move from dyadic interactions to localized but coalitional conflicts to regional or global conflagrations of greater scope. Just as every bilateral war begins as a crisis not too dissimilar from others that do not escalate to war, every world war has equally humble beginnings, and we can learn a great deal about conflict expansion by comparing those crises that remain localized to those that do not.

While the puzzle of alignment, balancing, and conflict expansion is hardly new, the question of the durability of victorious coalitions has gone largely unexplored by students of international relations, and Chapter 5 also generates predictions over this final phase in the life span of coalitions; that is, their ability to cooperate in the maintenance of the new status quo, the division of the spoils of victory, after winning a war together. Since postwar durability plays such a critical role in this chapter's theory, I also analyze a novel empirical model of the durability of war-winning coalitions and show, as expected, that the diversity of preferences plays a significant role in shaping the time between victory and the eruption of new conflicts between former partners – further increasing confidence in the utility of the theoretical model and indicating the fruitfulness of further exploration of the duration of postwar cooperation.

2.2.4 Summary

Coalitions are distinguished from states acting alone by their aggregate military power and the internal bargains over the terms of cooperation that balance the diversity of preferences within them. This places a premium on explaining how coalitions form, when and why their crises escalate to war, and the conditions under which they provoke balancing behavior on their targets' behalf. The approach sketched earlier in the chapter offers a rationale for tracing the effects of both power and preferences through the entire process, highlighting as it does the critical role of securing and preserving military cooperation from one's partners. At the stage of coalition formation, a state weighs the costs of securing cooperation against the military benefits of receiving it in forming a coalition. In the crisis bargaining stage, the content of those compromises – in the case of Chapter 4, altering escalation levels to accommodate nervous partners – has some heretofore unexamined consequences for the credibility of signals and the probability of war. Specifically, coalitions send more credible signals of resolve against stronger targets because they are discouraged from bluffing lest they lose coalitional support; on the other hand, they are more likely to send "weak" and ineffective signals against weaker targets. Finally, in the third stage, conflict expansion, the quality of military cooperation as reflected in a coalition's durability also interacts with its aggregate power to influence the decisions of outsiders not originally involved to enter coalitional conflicts and expand them still further; powerful coalitions can discourage balancing when their interests are diverse, but the same heterogeneity is a liability in terms of provoking balancing for weaker coalitions. Finally, the same diversity of preferences that necessitated bargaining and compromise in their formation can also shape coalitions' survival long after they have defeated their targets, and in the following section I introduce the data collected in order to understand these processes in greater detail.

2.3 INTRODUCING THE COALITIONS DATA

Despite the prevalence of military coalitions, political scientists have produced little systematically collected data on their makeup, characteristics, and participation in international crises (Snyder 1991, p. 132). Tago's (2007) study of coalition formation is a notable exception, although the data focus exclusively on explaining why states join American-led military (and some humanitarian) efforts, while Starr (1972) and Morey (2011) examine wartime coalitions only, differing in the extent to which they require coordinated strategies on the battlefield.[31] Thus, to test the implications of the models developed in subsequent chapters, I collected data on coalition formation and behavior in a sample of interstate crises as identified by the International Crisis Behavior Project for the 1946–2001 period (Wilkenfeld and Brecher 2010).[32] In this section I give a brief overview of the data and their relationship to the definition and concepts outlined in the preceding two sections, and then leverage them to uncover the bivariate links between coalitions and (a) the probability of war and (b) conflict expansion, establishing some empirical patterns and facts to be explained by the models in Chapters 3–5.

2.3.1 Identifying Coalitions

Focusing on coalitions at the level of the individual crisis carries with it some significant advantages over existing approaches to military cooperation. First, crises are useful for identifying a large number of discrete cases in which military cooperation (coalition formation) and its absence (singleton disputants) are (somewhat) easy to define, observe, and measure. This has the added benefit of facilitating measurement of the consequences of cooperation, from whether crises escalate to war to whether they provoke third parties to balance against them. In contrast, the literature on great power threats and balancing draws on only a limited number of observations (e.g., Mearsheimer 2001, Schweller 1994, Walt 1987). Additionally, while studies of the quality of allied cooperation use larger datasets to explore abrogation rates (Gaubatz 1996, Leeds and Savun 2007), wartime reliability (Leeds 2003a), and general deterrence (Johnson and Leeds 2011, Leeds 2003b), a focus on coalition-building in crises expands our understanding of military cooperation to conflict-specific instances of its occurrence, whether or not alliances are invoked or allies are involved. Finally, crises are defined in part by the possibility of "military hostilities" (Brecher and Wilkenfeld 1997, p. 3), making it easy to distinguish between military and diplomatic multilateralism and providing an empirical clarity that facilities the identification and separation of their effects.

---

[31] Kreps (2011), as noted earlier, samples on American-led "interventions," which carry a specific definition more restrictive than my focus on crises, and identifies coalitions with the presence of military and diplomatic multilateralism.

[32] I rely on ICB data to identify disputes, as opposed to the Correlates of War Project's Militarized Interstate Dispute data (Ghosn, Palmer, and Bremer 2004), because of its strict focus on the perceptions of national decision makers, which excludes a number of incidents – say, accidents or unauthorized border crossings – that do fit the definition of disputes in the MID data.

The International Crisis Behavior (ICB) Project's data take as their unit of observation the international crisis, defined by three conditions involving the beliefs

held by the highest level decision makers of the state actor concerned: a *threat to one or more basic values*, along with an awareness of *finite time for response* to the value threat, and a *heightened probability of involvement in military hostilities* (Brecher and Wilkenfeld 1997, p. 3, emphasis in the original).

Thus, international crises find state leaders contending with other states over scarce goods or "values," such as territory, policy, domestic institutions, rights and privileges, political influence, and so on. Disagreements over the distribution of these goods become crises when leaders believe that either they or their opponents may need to resolve them with military force, hence the focus on finite times of response – often associated with demands for change of the status quo – and an increased probability of war. As such, the ICB data facilitate a focus on two-sided interstate crises in the shadow of possible war, resulting in a sample of 261 crises from 1946 to 2001.[33] The data also identify whether the crisis escalated to war according to the ICB, as well as the states on each side, which serves as the starting point for coding the existence, membership, and characteristics of military coalitions. Thus, the primary unit of analysis in the ensuing chapters will be the crisis, since it is defined by the shadow of possible war, but only some of those involve possible wars in which states will fight alongside one another.[34]

After identifying which states participated on which side in each ICB crisis, identifying coalitions involves four steps. First, if ICB identifies multiple states on a side as originators or "triggering entities" of the crisis, then a side is deemed a possible coalition.[35] Second, to distinguish between military and

---

[33] While more ICB crises occurred during this period, intra war crises (Brecher and Wilkenfeld 1997) of the Korean, Vietnam, and Iran-Iraq wars are collapsed into a single crisis for the original participants. One-sided crises and those in which target states are already involved in war are dropped, as well as several crises involving states for which various data are unavailable: Iceland (twice), the Solomons, Malta, and (depending on the relevant data) Grenada.

[34] Crises are triggered when states perceive threats to key values, and these initial perceptions can derive from explicit demands for change, such as Soviet threats to sign a peace treaty with East Germany in 1961 (ICB #185), or implicit threats, such as China's incursion across the McMahon line and into Indian territory in 1962 (ICB #194). Some threats are backed up by threats of coalitional action (the former) while others are unilateral (the latter), and since (a) coalitions are defined by the perception of such threats and (b) states choose their coalition partners strategically, we might expect that the observation of a crisis and a coalition may in some cases not be entirely independent of one another. To address this, I model these dependencies explicitly in the following chapters' theoretical models and estimate a model of coalition formation with sample selection to assess the consequences of treating crises and the coalitions that participate in them as independent events.

[35] A triggering entity is the state that "initiated the act which was perceived by a state as creating a threat to basic values, time pressure and heightened probability of military hostilities" (Wilkenfeld and Brecher 2010).

diplomatic multilateralism, any potential coalition member must also make or be expected by other states to make a military contribution, either deploying military forces or providing basing or staging assistance. The expectation of military contributions is key, since not all crises escalate to war; coding only for observed military participation would censor those threats of collective military action that successfully averted hostilities. Third, if a state is coded by ICB as participating on a side but joins only after the crisis escalates to war, it is coded *not* as an original coalition member but as a joiner in a potential balancing or counter-coalition scenario. Finally, if reading the crisis summaries provided by ICB and other secondary sources indicates that members of an ostensible coalition were not understood to be threatening or considering collective military action, then they are *not* coded as coalition members, overruling any prior indication to the contrary according to the first three rules. For each coalition, this results in a list of members that is narrower than the more inclusive ICB definition of crisis participation, given my interest in explicitly military modes of cooperation.

Several brief examples clarify how the coding rules work in practice. First, France, Israel, and the United Kingdom constitute a coalition in their attempt to seize the Suez Canal in 1956 (ICB #152), but threats of involvement by the Soviet Union once hostilities commenced do not count as a coalition forming around Egypt, nor does the American attempt to restrain its allies constitute coalition membership on its part. However, as discussed in Chapter 5, the Soviet Union would be counted as siding with Egypt in response to the threat posed by the initial coalition, balancing against it by siding with its target. Second, Sudan and Afghanistan, the twin targets of American cruise missile strikes in response to the 1998 embassy bombings in Kenya and Tanzania, are coded as on the same side of ICB crisis #427. However, since there is no evidence of coordination on their part either in the bombings that precipitated the strikes or in a response to the strikes, they do not constitute a coalition. In fact, given the lack of any coordination on the part of the Sudanese and Afghan governments in this crisis, I disaggregate it into two separate crises. By contrast, Zambia and Angola did form a coalition after being separately targeted by Rhodesia in cross-border raids in 1979 (ICB #300), as they issued a joint threat to retaliate in the event of further attacks launched from Rhodesia.

Finally, the threat of military participation, as opposed to other types of involvement, is critical. Even though ICB identifies Taiwan as a member of the United States–South Korean coalition in the Korean War (ICB #132), the American decision to neutralize the island (Stueck 1995, 2004), preventing it from engaging in hostilities against mainland China by deploying a naval task force to the Taiwan Strait, means that it does not meet the standard of coalition member in these data. Further, even when states work together explicitly to shape the outcome of crises, they must wield an implicit or explicit threat to use force. Thus, the United States and the United Kingdom's attempts to mediate between Italy and Yugoslavia during their 1953 crisis over Trieste (ICB #142) is not a military coalition, despite the obviously high level of cooperation.

For each crisis, I also identify the principal belligerents, or those two states that would be involved in the substance of the dispute absent the presence of others.[36] In other words, principal belligerents are potential coalition-builders. I do this by identifying the primary issue of the crisis – a disputed border, support for transnational insurgents, a regime to be changed – and then coding the two states contending over that primary issue. For dyadic crises, this is straightforward. For coalitional crises, I consult ICB crisis summaries to identify which pair of states experience the first exchange of coercive threats or attempts to revise the status quo over that issue. For example, the principal dyad in the Korean War (ICB #132) is North and South Korea, since the former invades the latter; in the Cuban Missile Crisis (ICB #196), the United States and the Soviet Union, since the former demands policy changes of the latter after the discovery of missile installations in Cuba; and in the Suez Crisis (ICB #152), the United Kingdom and Egypt, since the latter objected to the continuing presence of the former's troops in the Canal Zone and abrogation of commitments to support construction of the Aswan High Dam. In those cases for which there is a disparity between parties to the principal dispute and the issuance of the first threat, I code principal belligerents as those exchanging the first threats. For example, in the Gulf of Tonkin crisis (ICB #210), the primary issue is between North and South Vietnam, but the initial coercive moves occur between North Vietnam and the United States, and I identify these two as the primary belligerents.

2.3.2 The Characteristics of Coalitions

The resulting list of seventy coalitions and their principal belligerent dyads can be found in Table 2.7 in this chapter's appendix, but at this point a few preliminary summary statistics are revealing.[37] First, coalitions appear on at least one side of a crisis in roughly 25% of the observations in the sample. The modal coalition has two members, as shown in Figure 2.2. There are a smaller number of medium-sized coalitions with three to five members, but the two largest – associated with the Gulf War of 1991 and the Kosovo War of 1999 – produce a rather skewed distribution. Notably, according to Keohane's (1990) definition of multilateralism, which requires at least three states, the majority of coalitions in this sample would be eliminated from the analysis. Second, as shown in Figure 2.3, coalitions have appeared with impressive regularity throughout the sample period, averaging around one-and-a-half per year, with a maximum of five coalitions in 1979. There are noticeable peaks of coalitional

---

[36] This is equivalent to the definition of "lead state" that I use in the models found in Chapters 3–5.

[37] Principal belligerents are italicized in Table 2.7.

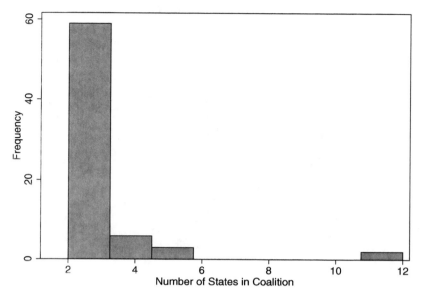

FIGURE 2.2. Histogram of observed coalition sizes, 1946–2001

activity (a) in the late 1950s–early 1960s, at the heights both Cold War and Arab-Israeli tensions; (b) in the late 1970s, dealing mostly with crises in Africa and South Asia; and (c) in the 1990s, as the United States led coalitions in the Middle East and in the states of the former Yugoslavia.

Turning to a comparison with states acting alone, or singletons, Table 2.2 shows that coalitions differ in some notable ways, even as the two share some interesting similarities. Unsurprisingly, coalitions have roughly six times the military capabilities of singletons, as measured by the Correlates of War Project's Composite Index of National Military Capabilities (CINC: Singer, Bremer, and Stuckey 1972, Singer 1987); in fact, the average coalition is made up of about 12% of all the military capabilities in the international system in a given year, lending some plausibility to the idea (central to Chapter 5) that third parties view them as potentially more threatening for the future than singleton belligerents. However, their targets are not statistically stronger in terms of military capabilities than the targets of singletons, differing by an average of only 0.01 in their CINC scores.[38] As a result, coalitions also account for a far larger share of the total capabilities in their crises, controlling nearly 80% of the crisis-level distribution of power as opposed to roughly half for singletons. Thus, coalitions tend to be more powerful than singletons in both absolute and

---

[38] A two-sample t-test fails to reject the null hypothesis that target capabilities for coalitions and single disputants are drawn from the same population ($p < 0.001$).

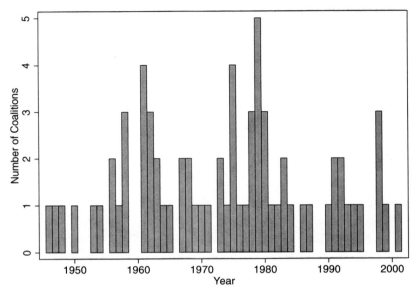

FIGURE 2.3. Number of coalitions by year, 1946–2001

relative terms, which, as discussed later, may go some length toward explaining their unique behavioral patterns.

Coalitions are also, on average, no more democratic than singletons; the average Polity-IV combined democracy score (Marshall and Jaggers 2009) inside coalitions is statistically indistinguishable from the democracy score of the average singleton that is a mixed regime that leans, if slightly, closer to autocracy than democracy.[39] Additional data discussed at greater length in Chapters 3 through 5 show that coalitions are also more likely than singletons to have the explicit support of the United Nations Security Council; coalitions receive such support roughly 11% of the time as opposed to a more modest 3% rate of UNSC support for singletons.[40] Finally, despite their advantages in observable military capabilities, the median coalitional crisis is roughly three times longer that that of the median bilateral crisis (Wilkenfeld and Brecher 2010), which suggests that they may be costlier and/or more difficult to resolve – or, alternatively, that coalitions tend to form around the issues most likely to result in protracted conflicts.

Aggregating military power has long been recognized as a popular motivation for military cooperation, and Table 2.2 confirms that building coalitions does, on average, produce greater relative power in crises. However, military

---

[39] Mixed regimes, or "anocracies" are generally considered to fall between −7 and 7 on the twenty-one point (-10,10) combined democracy scale (Marshall and Jaggers 2009).

[40] As discussed in Chapter 3, this pattern may emerge due to the positive effect of UN support on states' willingness to join coalitions.

TABLE 2.2. *Comparing Coalitions and Singletons, 1946–2001*

Variable	Singleton	Coalition
Military capabilities (COW)	0.02	0.12
Target capabilities (COW)	0.02	0.03
Relative capabilities (COW)	0.50	0.80
Democracy (Polity-IV)	−0.85	−0.86
Pr(UNSC support)	0.03	0.11
Median Crisis Duration in Days (ICB)	60	193

Data from Singer, Bremer, and Stuckey (1972), Marshall and Jaggers (2009), and Wolford (2014a).

power also plays a role in decisions over war and peace, as well as balancing and remaining neutral, which raises the final question of the chapter. If coalitions *do* exhibit different rates of crisis escalation and/or conflict expansion than singletons, how much can be attributed simply to their uniquely large amounts of military power?

2.3.3 Crisis Escalation and Conflict Expansion

If observed differences in crisis escalation and conflict expansion between coalitions and singletons can be explained with the same set of factors – for example, with reference to their uniquely high levels of military capabilities – then seeking to explain coalitional behavior would be a trivial exercise. I show in this section, however, that this is not the case. I first use simple cross-tabulations to establish that crises involving coalitions do exhibit unique patterns of escalation and expansion, then conduct a more sophisticated decomposition analysis to show that simple differences in aggregate and relative power can account for only a very limited share of the differences between coalitions and singletons (Blinder 1973, Fairlie 2005, Oaxaca 1973).[41] Thus, in addition to laying out some of the facts to be explained, these analyses also show that coalitions do not behave simply as very powerful singletons; they are, in fact, something more than the sum of their parts, which highlights the need for the coalition-specific theoretical models and concepts developed in subsequent chapters.

Starting with simple bivariate relationships, Table 2.3 shows that crises involving coalitions on at least one side are substantially more likely to escalate to war than those with singletons.[42] When states enter crises without partners, the probability of war is 0.11, but the probability jumps to 0.37 when coalition partners are involved, and this association is statistically discernible

[41] For a recent application of decomposition models in political science, see Reed and Chiba (2010).
[42] Each observation in Tables 2.3 and 2.4 is a "side" in a directed ICB crisis, such that the total number of observations is twice the number of crises in the sample.

TABLE 2.3. Coalitions and Crisis Escalation, 1946–2001

	Escalates to War		
Coalition	No	Yes	Total
No	404	48	452
Yes	44	26	70
Total	448	74	522

$\chi^2_{(1)} = 35.05, p = 0.0001$

at $p < 0.0001$. In other words, the probability that this difference would be observed by chance is less than one-hundredth of one percent. Likewise, in Table 2.4, coalitions are also significantly ($p < 0.009$) more likely to provoke balancing than singletons, a difference in probability between 0.13 and 0.24. What accounts for these differences? One possibility, of course, is that coalitions are simply more powerful than singletons, prompting them to (a) underestimate weaker opponents' reputation costs and make demands likely to be rejected (see Sechser 2010) or (b) pose such substantial threats to third parties that they provoke balancing in the form of counter-coalitions (Walt 1987, Waltz 1979). By each of these lines of argument, coalitions are no different than equivalently powerful single disputants. Put simply, they are pure aggregations of power – no more, no less – and subject strictly to the same behavioral incentives and constraints appropriate to their level of capabilities.

One way to judge whether coalitions can appropriately be treated simply as very strong singletons is to estimate an empirical model that decomposes the effects of military capabilities across coalitions and singletons into observable and unobservable components, then assess how much of the difference between groups can be attributed to different levels of military power and to unobserved factors, such as preference diversity, omitted from the analysis. I conduct these analyses using Fairlie's (2005) extension of the Blinder-Oaxaca decomposition technique (Blinder 1973, Oaxaca 1973) to the special case of binary response

TABLE 2.4. Coalitions and Conflict Expansion, 1946–2001

	Provokes Balancing		
Coalition	No	Yes	Total
No	395	57	452
Yes	53	17	70
Total	448	74	522

$\chi^2_{(1)} = 6.79, p = 0.009$

models. The models in Tables 2.5 and 2.6 decompose the difference in outcome probabilities – that is, escalation and expansion across coalitions and singletons – into fractions explained by "endowments," or different levels of military power, and "coefficients," or different behavioral responses at the same values of the variables. Formally, a decomposition model tells us, first, which variables make a statistically discernible difference in the outcome probabilities and, second, the share of the difference for which observed variables (endowments) account. Intuitively, the test assesses whether coalitions behave mostly like singletons with the same military capabilities or whether other, unmeasured factors account for differing behavioral responses given the same level of capabilities.

Table 2.5 presents the results of two empirical models: first, a simple probit model of the escalation of crises to war and, second, the decomposition analysis showing the contribution made by each military power variable to explaining the difference in the rates of escalation between singletons and coalitions. The sample for each is a set of directed-dyadic crisis-sides; that is, a given crisis is observed twice, with each side having the chance to be observed as side 1 and again as side 2. For each observation, I include an indicator of whether side 1 included a coalition. I also include a measure of the aggregate military capabilities of each side as defined by the Correlates of War Composite Index of National Capabilities (Singer 1987, Singer, Bremer, and Stuckey 1972), side 1's share of the total capabilities in the crisis to capture differences in relative power, and a control for the number of states also participating on side 2. Model 1 presents the results of estimating a probit model of the probability that the crisis escalates to war, where the presence of a coalition on side 1 is associated with a statistically discernible ($p < 0.01$) increase in the probability that the crisis escalates to war, even in the presence for controls related to military capabilities and the number of states on side 2.

The probit model does not, however, rule out entirely the possibility that coalitions appear distinct because of their uniquely large endowments of military capabilities. The decomposition analysis in Model 2 offers a way to assess this by using the estimates of the probit model to compare coalitions and singletons along two dimensions: their values of the explanatory variables and how they respond to those variables. In other words, a decomposition analysis can indicate what proportion of the difference in the predicted probability of war, 0.11 for singletons and 0.37 for coalitions, is due to differences in military capabilities and what proportion is due to unobserved or unmodeled factors that would explain coalitions' differing ways of using or responding to military capabilities (see Fairlie 2005, pp. 306–309). The upper rows in Model 2 show, first, that none of the measures of military capabilities, whether absolute or relative, can explain a significant portion of the differences between coalitions and singletons. This suggests that coalitions are, indeed, something more than the sum of their parts. In fact, as indicated by the tiny share of the total difference explained by endowments, $(-0.03/-0.26) \approx 0.11$, the vast majority of

TABLE 2.5. *Decomposition Analysis of Coalitions and Singletons in Crisis Escalation, 1946–2001*

	Pr(War = 1)	
Variable	Model 1 *Probit*	Model 2 *Decomposition*
Coalition$_1$	1.20 (0.21)***	–
CINC$_1$	−2.30 (1.31)*	−0.02 (0.03)
CINC$_2$	−2.26 (1.52)	0.01 (0.01)
Relative Capabilities	−0.19 (0.28)	−0.001 (0.02)
Number$_2$	0.32 (0.13)**	−0.02 (0.01)
Intercept	−1.47 (0.22)***	–
Pr(War|No Coalition)		0.11
Pr(War|Coalition)		0.37
Difference		−0.26
Total Explained		−0.03

Significance levels: * : 10%   ** : 5%   *** : 1%

differences between coalitions and singletons appears to be explained by other factors – specifically, those that might explain why coalitions behave differently than singletons at the same levels of military power.

Turning next to the question of conflict expansion, where crises grow to include countries beyond their original disputants, Table 2.6 presents a similar

TABLE 2.6. *Decomposition Analysis of Coalitions and Singletons in Crisis Expansion, 1946–2001*

	Pr(Expand = 1)	
Variable	Model 1 *Probit*	Model 2 *Decomposition*
Coalition$_1$	0.48 (0.28)*	–
CINC$_1$	−1.53 (1.32)	0.01 (0.04)
CINC$_T$	7.64 (1.64)***	−0.01 (0.01)**
Relative Capabilities	0.66 (0.30)**	−0.05 (0.02)**
Target Borders	−0.06 (0.03)***	0.02 (0.01)
Intercept	−1.42 (0.28)***	–
Pr(War|No Coalition)		0.13
Pr(War|Coalition)		0.24
Difference		−0.11
Total Explained		−0.03

Significance levels: * : 10%   ** : 5%   *** : 1%

pairing of a simple probit model with a coalition on side 1 as a key explanatory variable and a decomposition analysis of the differences between coalitions and singletons. In this case, however, the outcome variable, discussed at greater length in Chapter 5, indicates whether the target state on side 2 received third-party assistance during the conflict; that is, whether states not party to the original dispute entered or threatened to enter the conflict militarily in opposition to side 1. The probit estimates show that coalitions are, indeed, significantly more likely to provoke balancing than singletons ($p < 0.1$), controlling again for the absolute and relative capabilities of side 1 and its original target, as well as the number of states bordering the target.

Moving to the decomposition analysis in Model 2, two things are of note. First, both side 2's aggregate capabilities and relative capabilities – roughly side 1's chances of defeating its target absent any balancing – make a significant contribution to explaining part of the difference. In fact, differences in capabilities account for roughly 27% ($-0.03/-0.11$) of the difference in singletons' and coalitions' rates of conflict expansion, leaving about three-fourths of the difference to be explained by the inclusion of underlying factors driving these behavioral differences. Therefore, in the realm of conflict expansion as well as crisis escalations, coalitions cannot usefully be understood as mere aggregations of power; they appear to be more than the sum of their parts. As I argue in the following chapters, the politics of their formation, bargaining strategies, and expected durability are all central to explaining the underlying differences between coalitions and states acting alone in crises.

2.4 SUMMARY AND DISCUSSION

This chapter has argued that coalitions are analytically distinct from other, more commonly studied forms of international military cooperation. Military coalitions are neither alliances nor diplomatic coalitions, and the traditional ambiguity of the concept makes them difficult to define. In a nod to the insufficiency of treaties of alliance to capture many forms of cooperation relevant to the puzzle of war and peace, Snyder (1991) noted a quarter-century ago that "the notion of 'coalition' does not have a clear-cut empirical analogue," identifying the contours of the basic empirical problem that this chapter seeks to address:

In international life . . . the existence of a coalition does not depend on formal agreement. Mutual expectations of military support may follow from a variety of behaviors short of formal alliance or merely from a coincidence of interests. (p. 132)

This ambiguity about what constitutes a coalition has been difficult to overcome – the proliferation of definitions in the literature itself is testament to this – especially in the face of obvious alternatives like treaties of alliance and coalitions that confront revisionist great powers. However, while the former are neither necessary nor sufficient to produce coalitions, the latter are only a subset of the broader, indeed much richer, category of international military coalitions.

In this chapter, I offer one solution to the conceptual ambiguity surrounding coalitions by focusing on a particular type that sits somewhere in between these two extremes. To be sure, my definition has limitations; it excludes some groupings that we might call coalitions, say if one state provides only financial support to another, but its restrictiveness is a strength in that it makes the presence of coalitions in discrete international crises easy to identify and observe. As such, I have put forward a clear definition and identified what appear to be two of its important properties: (a) substantial military power (relative to singletons) and (b) a diversity of preferences within the membership. I have argued that these two concepts may prove useful in understanding all three phases of military multilateralism, from the formation of coalitions, through the escalation of their crises to war, to the provocation of balancing. Using my operational definition of coalitions, I coded data on coalition participation in a sample of late twentieth and early twenty-first century international crises. Finally, a pair of decomposition analyses indicates that coalitions contribute something unique, something beyond their considerable military capabilities, to the processes of crisis escalation and expansion.

Though I generate a number of testable implications from the theoretical models in the following chapters, I focus on only a few dependent variables: (a) which, if any, partners that potential coalition-builders choose to form coalitions with; (b) the escalation of crises to war; (c) the occurrence of third-party balancing on behalf of a target state; and (d) the durability of cooperation among victorious coalition members after war. Crisis escalation and expansion, in particular, require explanations different than those typically advanced for states acting alone. Each of these outcomes are of enduring interest in their own right to scholars of international relations, as evidenced by the substantial literatures on alliance formation, crisis bargaining, and the causes of war, as well as alignment decisions and the expansion of conflicts. The theoretical and empirical models in this volume represent an initial attempt to develop a common set of answers to these questions derived from the politics of crisis-specific military coalitions – important yet understudied phenomena in their own right.

2.5 APPENDIX

Table 2.7 presents each crisis for which coalition participation was identified. Italics denote primary belligerents, and partners appear below the primary belligerent on whose side they joined the crisis.

TABLE 2.7. *ICB Crises Involving Coalitions, 1946–2001*

ICB #	Year	Side 1	Side 2
111	1946	*Soviet Union*	*Turkey* United States United Kingdom
120	1947	*Egypt* Iraq Syria Lebanon Jordan	*Israel*
123	1948	*Soviet Union*	*United States* United Kingdom France
132	1950	*North Korea*	*South Korea* United States
144	1953	*Guatemala*	*Honduras* United States
146	1954	*China*	*Taiwan* United States
152	1956	*United Kingdom* France Israel	*Egypt*
153	1956	*Israel*	*Jordan* Iraq
159	1957	*Turkey* United States	*Syria*
166	1958	*China*	*Taiwan* United States
168	1958	*Soviet Union* East Germany	*United States* United Kingdom France West Germany
180	1961	*United States* Thailand	*Vietnam*
183	1961	*Iraq*	*Kuwait* United Kingdom
185	1961	*Soviet Union* East Germany	*United States* United Kingdom France West Germany
195	1962	*Saudi Arabia* Jordan	*Yemen* Egypt

(*continued*)

TABLE 2.7 *(continued)*

ICB #	Year	Side 1	Side 2
196	1962	*Soviet Union* Cuba	*United States*
202	1963	*Cyprus* Greece	*Turkey*
203	1963	*Egypt* Syria Lebanon Jordan	*Israel*
210	1964	*United States* South Vietnam	*North Vietnam*
216	1965	*Pakistan* China	*India*
222	1967	*Egypt* Syria Jordan	*Israel*
223	1967	*Cyprus* Greece	*Turkey*
224	1968	*North Korea*	*United States* South Korea
227	1968	*Soviet Union* East Germany Poland Hungary Bulgaria	*Czechoslovakia*
232	1969	*Egypt* Soviet Union	*Israel*
238	1970	*Syria*	*Jordan* United States Israel
242	1971	*Bangladesh* India	*Pakistan*
255	1973	*Egypt* Soviet Union Syria	*Israel* United States
257	1974	*Cyprus* Greece	*Turkey*
260	1975	*Zaire* United States Zambia South Africa	*Angola* Cuba Soviet Union

ICB #	Year	Side 1	Side 2
261	1975	*Spain* Algeria	*Morocco* Mauritania
274	1976	*North Korea*	*United States* South Korea
284	1977	*Vietnam* Thailand	*Cambodia*
290	1978	*Chad* France	*Libya*
292	1978	*Zaire* United States Belgium France	*Angola*
296	1978	*Uganda* Libya	*Tanzania*
300	1979	*Rhodesia*	*Zambia* Angola
303	1979	*Soviet Union* Afghanistan	*Pakistan*
304	1979	*Chad* France	*Libya*
306	1979	*Soviet Union* Afghanistan	*Pakistan*
307	1979	*Rhodesia*	*Botswana* Mozambique Zambia
311	1980	*Tunisia* France	*Libya*
315	1980	*Soviet Union* East Germany Czechoslovakia	*Poland*
320	1980	*Ethiopia* Kenya	*Somalia*
334	1981	*Iran*	*Bahrain* Saudi Arabia
337	1982	*Israel*	*Lebanon* Syria
340	1983	*Libya*	*Sudan* Egypt
342	1983	*Chad* France	*Libya*

(continued)

TABLE 2.7 *(continued)*

ICB #	Year	Side 1	Side 2
350	1984	*Libya*	*Sudan* Egypt
362	1986	*Chad* France	*Libya*
375	1987	*Algeria* Mauritania	*Morocco*
393	1990	*Iraq*	*Kuwait* United States United Kingdom France Egypt Syria Saudi Arabia Bahrain Qatar United Arab Emirates Oman Turkey
397	1991	*Serbia*	*Slovenia* Croatia
399	1991	*France* Belgium	*Zaire*
403	1992	*Serbia*	*Bosnia* Croatia
406	1992	*United States* United Kingdom France	*Iraq*
408	1993	*North Korea*	*South Korea* United States
412	1994	*Iraq*	*Kuwait* United States Saudi Arabia
415	1995	*China*	*Taiwan* United States
426	1998	*Zaire* Chad Angola Zimbabwe Namibia	*Rwanda* Uganda

ICB #	Year	Side 1	Side 2
429	1998	*United States* United Kingdom Kuwait Oman	*Iraq*
430	1999	*United States* Canada United Kingdom Netherlands Belgium France Spain Portugal Germany Italy Albania	*Serbia*
434	2001	*United States* United Kingdom	*Afghanistan*

3

Power, Preferences, and Cooperation

> There was no prospect of victory over Germany, unless it were with the help of Russia... Their military purposes, in other words, were mortgaged in advance.
> George F. Kennan
> *American Diplomacy*, p. 77

In 1991 and 2003, the United States led coalitional wars against Iraq that relied on military cooperation, particularly basing and staging rights, from nearby states. Iraq's northern neighbor, Turkey, figured prominently in prewar coalitional negotiations in each case, as the United States was attracted by the potential to threaten Iraq with a two-front war. Turkey allowed its territory to be used to stage part of the massive air campaign preceding the invasion of Kuwait and southern Iraq in 1991, not to mention tying down a large part of the Iraqi army with the threat of opening a northern front; however, in 2003, its parliament refused an American offer to join the coalition that ultimately toppled Iraq's Ba'athist government. Turkey's potential military value to the United States was significant in each war, primarily centered on useful basing, staging, and transit areas. It was also tied to the United States through a formal alliance commitment (NATO), and it shared American concerns about preserving order in the region and the creation of a stable post-Ba'athist Iraq. What, then, accounts for Turkey's participation in 1991 and its refusal to do so in 2003? Part of the answer, to be sure, is Turkey's negative reaction – not uncommon among the NATO allies – to escalating American aims across the two wars. Yet, as discussed later in the chapter, the United States still might have been able to secure Turkey's cooperation in 2003. Nonetheless, it failed. Why?

In this chapter, I develop and test an answer to this question by focusing on the strategic interaction between a would-be coalition-builder and a potential partner, who must be compensated for its cooperation in the costly

endeavor of crisis bargaining and possible war. Whether it occurs within legislatures (Baron and Ferejohn 1989), among rebel groups challenging the same
state (Akcinaroglu 2012, Horne 2006), or between states that may fight wars
together (Morrow 1991, Palmer and Morgan 2010), building coalitions entails
a fundamental trade-off between increasing the chances of success and making political concessions to secure the cooperation of potential partners (Riker
1962). In fact, even after agreeing to a 60–40 split of their bilateral burden
after the invasion of Poland in 1939 (Hastings 2012, p. 27), British and French
leaders asked themselves during the gathering storm of the Second World War,

Was it worth making further efforts to persuade [the Soviet Union and the United States]
into an alliance against the fascist states, even if this involved substantial concessions to
Moscow's and Washington's requirements, and provoked criticism at home? (Kennedy
1989, p. 320)

Even for the highest of stakes – and in the Europe of 1939–1945, they could
scarcely have been higher – states devote considerable attention to the price they
must pay in return for any particular partner's assistance. How they resolve
this trade-off, whether through acting alone or taking on some partners and
not others, is the subject of this chapter. Later, Chapters 4 and 5 show that
a coalition-builder's choices in the face of this dilemma have implications for
the escalation of crises to war, the expansion of conflicts, and the durability
of cooperation after victory. First, however, it is necessary to understand the
underlying strategic dimensions of how such coalitions form in the fist place.
    As discussed in Chapter 2, a traditional focus on the politics of formal
alliances and, more recently, diplomatic multilateralism has left the underlying logic of and empirical patterns associated with coalition-building poorly
understood, despite the fact that coalitions are the dominant mode of military
cooperation in observed crises. To begin to fill this gap, this chapter proposes
and analyzes a stylized formal model of coalition formation that sheds light
on the linkages between power, preferences, and the political concessions that
secure military cooperation. I use the theoretical model to generate empirical
expectations over coalition partner choice that I then test against a sample of
late twentieth- and early twenty-first-century crises, showing that, while states
prefer to take on powerful partners with similar preferences, they become less
selective – that is, more willing to compensate partners with increasingly divergent preferences – when these partners offer larger military benefits and increase
the chances of ultimate success.
    In the remainder of this chapter, I present and analyze the basic theoretical
model of bargaining and coalition formation (a) from which I derive hypotheses
and (b) that informs the more sophisticated contributions to the modeling
dialogue in later chapters. While the theory treats events subsequent to coalition
formation, such as crisis bargaining and conflict expansion, in reduced form,
the general approach here ensures that these more complicated parts of the
process are easy to integrate more explicitly, and with a maximum of conceptual

overlap, into the models analyzed in Chapters 4 and 5. I focus the bulk of the discussion on the model's implications for coalition formation and partner choice, but I also derive additional implications for the initiation of crises and the relationship between the collective nature of the international stakes and the prospects for international cooperation. After deriving these implications, I analyze two empirical models, one of coalition partner choice and another of preference diversity inside coalitions. I then show how the model's logic explains why Turkey joined the U.S.-led coalition against Iraq in 1991 but did not, despite a more valuable military contribution than 1991 and more intensive negotiations, in 2003.[1]

3.1 THE PRICE OF COOPERATION

While most scholarship on military cooperation addresses either alliance formation (e.g., Gibler and Rider 2004, Gibler and Wolford 2006, Lai and Reiter 2000, Poast 2012, Siverson and Emmons 1991) or wartime collaboration (Choi 2004, 2012, Morey 2011, Reiter and Stam 2002), there is a small body of work on which to build in approaching the question of coalition-building during crises. As I argue in this section, though, it tends to explain coalition formation only in reference to the benefits of a particular partner or joint effort, often to the exclusion of the costs of cooperation, which are equally important in explaining which states build coalitions and what partners they choose.[2] Abstracting away from negotiations over the costs of cooperation has hindered the development of an understanding of the linkages between partner choice, crisis escalation, and conflict expansion, and this chapter is an attempt to begin to uncover those relationships.

Across several empirical domains and definitions of multilateralism, numerous studies have found associations between the attractiveness of either members or the mission itself and the formation of coalitions. Tago (2007), for example, shows that characteristics of the mission, including United Nations approval and the scope of its aims, are important predictors of which states are likely to join United States–led coalitions. Next, Vucetic (2011) uses the same data to show that English-speaking countries are also uniquely likely to join US-led coalitions. Still other studies have found that similar domestic institutions (Mousseau 1997), prior alliance commitments (Corbetta and Dixon 2004), and an interaction of the two (Pilster 2011) are all linked to military cooperation across both militarized interstate disputes and international

---

[1] As discussed further later, I also analyze a selection model in the appendix to assess the extent to which errors across the two stages of crisis onset and coalition formation are correlated. I find that, in fact, the errors are uncorrelated, increasing confidence that the empirical model in the main body of the chapter, which takes the onset of a crisis as given, is not contaminated by a damaging amount of selection bias.

[2] Kreps (2011), discussed in more detail later in the book, is a notable exception.

interventions. Finally, in an analysis of the politics of institutional authorization, Chapman (2011, ch. 5) shows that support from a relatively "conservative" UN Security Council – that is, one that should be ex ante unlikely to support a particular proposal – can also facilitate international cooperation, as more states tend to join the proposing state's side in an international crisis than would be the case without such approval.

The logic underlying each of these divergent theoretical and empirical models is simple: when a partner finds a mission or a coalition leader attractive, then a coalition is more likely to form. However, coalition partners come with varying price tags, some of which an ostensible coalition-builder may – the benefits offered by a potential partner notwithstanding – be unwilling to pay. As it lined up coalition partners for the war effort in Korea, for instance, the United States placed limits on partner contributions to preserve efficiency and political control of the war (Stueck 1995, pp. 56, 57), even turning down an offer of ground troops from Pakistan precisely because the requested costs of compensation were too high (p. 73). Turkey, likewise, both a democracy and an ally of the United States, famously refused to join the 2003 Iraq War coalition. To understand why lead states fail to build otherwise attractive coalitions, I argue that we must develop a better understanding of the costs entailed in securing cooperation.

In contrast to the work discussed earlier, Kreps (2011) acknowledges the costs of cooperation and posits a trade-off between the time required to negotiate and build coalitions and the immediacy of threats, especially for powerful countries like the United States with viable options for unilateral action – another factor often absent from studies of military cooperation.[3] However, by treating time as the main obstacle to coalition formation, this approach assumes that all partners *can* be compensated if only the coalition-builder is sufficiently patient.[4] Not all partners are created equal, however. Negotiations indeed take time, and crises impose serious temporal constraints on their participants by definition (Brecher and Wilkenfeld 1997). The *reason* they take time, though, is that lead states and potential partners bargain over the price of cooperation in tangible political goods – war aims, commitments to linked issues, spoils, and so on – and delay can be an effective negotiating tactic. In fact, potential partner states may charge higher prices when their contributions are more valuable, making coalition formation still more difficult, but also allowing for a better understanding of otherwise puzzling instances of failed coalition-building, like Turkey in 2003.

---

[3] Voeten's (2001) analysis of Security Council voting, where states choose between supporting or opposing a superpower in light of its ability to act outside the institution, is also based on the credibility of a dominant military power's outside option of going it alone.

[4] My theory makes no such assumption about the costs of cooperation. Further, absent a consistent empirical indicator of "time horizons" in a given crisis, incorporating these expectations into the empirical models below is prohibitively difficult. See also Wolford (2013).

Extant work does not address explicitly the processes of intra-coalitional bargaining by which goods are exchanged in return for military cooperation in international crises (cf. Morrow 1991, Riker 1962). Nonetheless, as discussed in Chapter 2, shares of the spoils of victory, territorial compensation, funding burdens, bargaining strategies, and war aims, as well as linked issues such as aid, trade, and defense commitments all enter into negotiations between potential coalition partners. Understanding how states bargain over these goods, when such negotiations succeed and when they fail is essential if we are to understand why lead states choose some partners over others. Indeed, since potential partners set their own price for cooperation, that price may increase in many of the factors that scholars have associated with attractive partners, such as military power or reliability. Further, by developing a better understanding of what goods, exactly, lead states pay their partners with, we can go on to link the initial choice of partner to the other stages of military multilateralism. As shown in Chapter 4, for example, the goods traded to secure cooperation are often bargaining strategies themselves – which have a direct impact on the probability of war.

3.2 A THEORY OF COALITION FORMATION

The theoretical model in this chapter is the simplest of the three in the book, highlighting the essentials of the transactional theory of coalition formation in as spare an environment as possible. One state, a potential coalition-builder or "lead" state, has the option of acting on its own against a target state in a crisis or proposing a coalition to a partner, who can either join the coalition or refuse to cooperate.[5] At the heart of this bargaining is the issue of compensation: What compromises over aims, linked policies, bargaining strategies, or the spoils of victory does the potential partner require in order to cooperate with the lead state? Coalitions offer increased chances of military success, but the central question of the chapter is how much the lead state might be willing to trade – what price it is willing to pay – to secure those benefits, given the partners available.

Analysis of the model focuses on the effects of several factors: the military capabilities of lead state, target, and potential partner, as well as the similarity of foreign policy preferences between leader and potential partner. Most of the discussion centers on three results that I analyze empirically. First, states prefer to take on coalition partners that share their foreign policy preferences, all else being equal, because such partners require less compensation; as a result, states

---

[5] By "lead state" I mean a primary disputant in a crisis, not necessarily that it is the most powerful member of a coalition; rather, it will participate in a crisis against the target whether or not it forms a coalition. It is not a "leader" in the sense of relaxing the unitary actor assumption, where national leaders must weigh both the public and their own private interests in foreign policy (see Chiozza and Goemans 2004).

also tend to avoid partners with divergent preferences. Second, lead states also find powerful partners more attractive than weak ones because they are more willing to compensate those that dramatically increase their chances of military success. Third, however, the effect of preference divergence is conditioned by a potential partner's military capabilities. Specifically, lead states are far less selective about a given partner's preferences when it offers a large enough boost to the chances of military success in the crisis, leading them to cooperate with states that, if they were less powerful, would be otherwise unattractive – that is, prohibitively expensive – partners.

I also derive some additional implications that I do not test empirically. First, the relative weight of the public and private stakes of the crisis – that is, the benefits a partner derives whether or not it participates and those it can receive only by cooperating – has differing effects on coalition formation depending on the size of the partner's military contribution. Specifically, an increasing public goods component makes weaker partners more attractive and stronger partners less attractive, suggesting that the very nature of the issues over which states contend can have an impact on their ability to secure military cooperation. Second, while the central strategic dynamic involves bargaining between coalition-builder and potential partner, requiring mutual consent to form a coalition, I also model the lead state's initial decision to initiate or enter a crisis, showing that the availability of coalition partners can encourage the onset of crises in the first place; in other words, the very possibility of military cooperation can increase the chances of international conflict, encouraging states to initiate crises when, if the only option were unilateral action, they would not. I also briefly consider the implications of this result for potential selection bias in models of partner choice, which informs a robustness check for selection bias in the appendix to the chapter.

After specifying the model formally, I discuss the equilibrium informally, saving the proofs of all propositions and technical statements for the appendix. In the process, I trace the links between the model's basic assumptions, the core concepts of power and preferences, and its conclusions before generating the hypotheses that will be tested subsequently against the historical record of early twentieth and late twenty-first-century international crises.

3.2.1 The Model

Suppose that a lead state L, a potential coalition-builder, has some policy dispute with a target state T, over which it can initiate an international crisis by seeking to change the status quo.[6] Should L initiate a crisis, it must then decide

---

[6] Note that "lead state" in this context simply means an original disputant or party to the political dispute, not the individual national leader of the state in question. It also implies nothing about power relations between leader and potential partner – that is, the lead state is simply the first participant on a given side, not necessarily the dominant state in the resulting coalition.

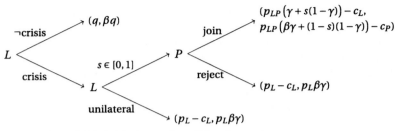

FIGURE 3.1. Crisis initiation and coalition formation

whether to propose a coalition to a potential partner, P, or to act unilaterally against the target state. Taking on a partner can improve the lead state's military prospects through capability aggregation or critical contributions like basing, staging, or transit rights. However, joining the coalition and contributing to the coercive effort is costly for the partner, who must be compensated with side payments to ensure its cooperation in the crisis and possible war.[7]

Before discussing the sequence of play, it is worth describing how I model the stakes of the dispute. Specifically, the stakes – territory, influence, policy, rights and prerogatives, and so on – have both a public (non-excludable, non-rival) and a private (excludable, rival) component. Normalizing the stakes to a total value of 1, both L and P have preferences over the substance of the policy outcome. For example, P may have its own preferences over the location of the lead state's border with the target or whatever policies the target might agree to change, regardless of whether P participates in the crisis. I represent the relative size of this public component with $\gamma \in (0, 1)$.[8] By refusing to participate in the crisis, P foregoes enjoyment of the other, private benefits that accrue to participants, denoted by $(1 - \gamma)$, such as control over aims and bargaining strategies, influence over the revised status quo, and even shares of the spoils of victory. This excludable element of the stakes is the share of the benefits that the lead state can keep for itself if it acts as a singleton but from which it may need to draw side payments in order to compensate a partner for its military contribution to a coalition.

As shown in Figure 3.1, the game begins with the lead state's choice over whether or not to initiate a crisis over the international status quo, which is

---

[7] While the sequence of moves allows coalitions to form only after a crisis begins, an alternative model might place coalition-building before crisis initiation. However, by placing formation second in the process, I ensure that the empirical implications do not rely on commitments made ex ante that, ex post, partners might not be willing to make. It also reflects the fact that the lead state is a principal belligerent, i.e., that it will participate in the crisis with the target regardless of whether it gains a partner or not.

[8] In an analysis of alliance formation, Benson (2012) treats defender and protégé preferences over the public component similarly. However, he abstracts away from the private component of the international issue, which is plausibly less important in alliance contracting than coalition formation.

simply a policy on the unit interval. The lead state values the status quo at $q \in [0, 1]$, and its ideal point is 1, while the target's ideal point is 0; this ensures that a genuine distributive conflict exists between the two. If L does not initiate a crisis, the status quo remains in place, and it receives simply $u_L(\neg\text{crisis}) = q$. For its part, P receives its own payoff for the status quo, $u_P(\neg\text{crisis}) = \beta q$, where $\beta \in [-1, 1]$ is a bias term representing the extent to which P shares L's foreign policy preferences; as β increases, P's preferred resolution of the issue at stake resembles L's (in opposition to the target's), but as β decreases, its preferences are increasingly opposed to L's and more in line with the target state's preferences.[9]

If L initiates a crisis in the first move, it then chooses whether to propose a coalition to P, which entails compromise, or to act unilaterally (that is, as a singleton). To isolate the processes of crisis onset and intra-coalitional bargaining, I treat the target as a nonstrategic player in this chapter, although it goes on to play a central strategic role in the signaling model of Chapter 4. To define payoffs for the crisis against the target, let Γ represent a general crisis game that resolves the crisis probabilistically in favor of either L or T, where the victor is allowed to set its preferred policy. Whether L acts unilaterally or with a partner, it pays a cost, $c_L > 0$, derived from military deployments and opportunity costs, as well as the destructiveness and expense of any conflict that may be fought to resolve it.

Next, since the outcome of the crisis is uncertain ex ante (cf. Morrow 1989, Wagner 2007), L succeeds in overturning the status quo with probability $p_L \in (0, 1)$ and fails with probability $1 - p_L$. I assume that L's probability of success, and therefore the size of its payoff, is correlated with its military power (see Banks 1990), which is consistent with the results of a broad number of crisis bargaining models (see, inter alia, Fearon 1995, Powell 1999).[10] Therefore, L's probability of success is tied to its military capabilities, $m_L > 0$, relative to the target's, $m_T > 0$, such that

$$p_L(m_L, m_T) \equiv \frac{m_L}{m_L + m_T},$$

where its probability of success is simply its share of the total military capabilities involved in the crisis. However, for the sake of clarity, I abuse notation and simply write p_L. Therefore, L's expected utility for a crisis in which it acts without a partner is simply $u_L(\Gamma, L) = p_L \times 1 + (1 - p_L) \times 0 - c_L = p_L - c_L$, where it pays the costs of the crisis regardless of the outcome, sets its ideal policy of 1 if it succeeds, and receives 0 if T succeeds and sets its own ideal policy. Notably, following a crisis in which it acts unilaterally, L enjoys both the public and private elements of success, which sum to 1 since $\gamma + (1 - \gamma) = 1$, while

---

[9] See Chapman and Wolford (2010) for a similar representation of preference similarity.
[10] For a similar representation of a reduced-form distributive contest, see Gilligan, Johns, and Rosendorff (2010).

P receives only its valuation of the public share of the policy outcome of the crisis, such that its expected payoff is $u_P(\Gamma, L) = \beta(p_L \times \gamma + (1 - p_L) \times 0) = p_L\beta\gamma$.

If, on the other hand, L wishes to take on a partner, it proposes a coalition to P that entails promising to transfer some of the private share of the stakes to P in return for cooperation.[11] L may limit its aims, demands, or escalation levels; engage in cheaper, less destructive, or lower-risk coercive strategies; share political influence over a successful outcome; or make more or less direct political or financial side payments, from formal defense commitments to aid packages. Whatever method of compensation it chooses, L sacrifices some of the value it would otherwise enjoy from a successful revision of the status quo, so I model its coalition proposal as some share of the private benefits of success. As such, if the coalition succeeds, L yields a share $s \in [0, 1]$ of the private stakes to P, such that the lead state keeps $s(1 - \gamma)$ and its partner gets $(1 - s)(1 - \gamma)$.

In return for the side payments made to ensure P's military cooperation, L's military prospects in the crisis improve, whether because it wields threats backed up by more military capabilities or because it wins a war if coercion fails; in either case, its payoff will increase in the total military capabilities it can bring to bear against the target. Formally, capability aggregation or force multiplication through basing or staging rights weakly increases L's chances of success, such that it succeeds in changing the status quo with probability $p_{LP} \geq p_L$. Thus, the coalition's probability of success is given by

$$p_{LP}(m_L, m_P, m_T) \equiv \frac{m_L + m_P}{m_L + m_P + m_T},$$

where $m_P \geq 0$ represents the partner's military contribution to the overall chances of success. As before, I abuse notation to ease presentation and simply write p_{LP} for the coalition's chances of success.

It is, of course, also plausible that burden- or cost-sharing might lower the costs of the crisis, even if P does not make a substantial military contribution. However, I focus here on capability aggregation because it makes for a stronger substantive match with the empirical models at the end of the chapter, which measure the value of a coalition partner in terms of its military capabilities. As such, L's costs for participating in the crisis are simply $c_L > 0$. Thus, should L secure P's cooperation, its expected utility for the crisis game is $u_L(\Gamma, LP) = p_{LP}(\gamma + s(1 - \gamma)) - c_L$. If P rejects the coalition, L acts unilaterally against the target, just as it would if it chose not to seek a coalition in the first place, and

---

[11] I assume throughout that this promise is credible. However, as evidenced by American-Soviet negotiations over the latter's entry into the war against Japan after Germany's defeat in Europe in 1945, partners and lead states both often doubt the other side's ability and willingness to follow through on their part of the bargain (Frank 2001). For further discussion of these issues, see Chapter 6.

payoffs are distributed accordingly.[12] For its part, P weighs the costs of joining the coalition against L's proposed compensation. If it rejects the coalition, it receives its payoff for a unilateral crisis, but if it joins the coalition, it receives $u_P(\Gamma, LP) = p_{LP}(\beta\gamma + (1-s)(1-\gamma)) - c_P$, where $c_P > 0$ represents its own costs for participating in the crisis as a member of L's coalition.

To summarize players' payoffs over terminal nodes, L's utility function is

$$u_L = \begin{cases} q & \text{if no crisis} \\ p_L - c_L & \text{if unilateral crisis} \\ p_{LP}(\gamma + s(1-\gamma)) - c_L & \text{if multilateral crisis,} \end{cases} \tag{3.1}$$

and P's utility function is

$$u_P = \begin{cases} \beta q & \text{if no crisis} \\ p_L \beta\gamma & \text{if unilateral crisis} \\ p_{LP}(\beta\gamma + (1-s)(1-\gamma)) - c_P & \text{if multilateral crisis.} \end{cases} \tag{3.2}$$

Equations (3.1) and (3.2) reveal the strategic tensions at the heart of building military coalitions and negotiating the terms of cooperation. For P, joining the coalition is costly, and it also may not value a resolution of the international issue in the same way as L; thus, it requires some compensation in order to make participation worthwhile. On the other hand, L would like to enhance its military prospects, but doing so comes at the cost of political concessions to a partner – limits on its aims, strategies, shares of the spoils, or political compensation – that it would rather not make. If a coalition is to form, then L and P must find some accommodation in the face of these competing interests. The following section examines the conditions under which each player's incentives are compatible in equilibrium, leading to the formation of a coalition, and when they are so incompatible that L instead opts to act alone.

3.2.2 Analysis

In this section, I answer two questions. First, once involved in a crisis, when will the lead state build a coalition, and when will it "go it alone" by acting unilaterally? Second, how does the availability of a coalition partner – that is, the promise of forming a coalition at an acceptable price – affect the lead state's willingness to initiate crises in the first place? The analysis produces several results of note. First, once a crisis has been initiated, lead states tend to build coalitions with states that share their foreign policy preferences, which ensures

---

[12] Failing to win such support can be costly, either diplomatically or domestically, especially when international institutions are involved (Chapman 2011, Chapman and Wolford 2010, Thompson 2006, Voeten 2005), but I omit such costs in order to isolate the effects of military cooperation incentives and distinguish them from any diplomatic/domestic reasons to seek out partners.

that L can purchase P's cooperation at an acceptable price. Second, lead states also seek out coalition partners that offer significant military benefits, whether because they need support against their target or because the partner offers a substantial military boost on its own. However, the more L benefits from P's cooperation militarily, the more it is willing to pay in return for cooperation, such that it is induced to tolerate partners with preferences increasingly divergent from its own. This factor turns out to be central in explaining the processes of both crisis escalation and conflict expansion in Chapters 4 and 5.

The model also yields to two other notable results, although they are not the main focus of the empirical model. First, the public share of the stakes of the crisis can make coalition formation either more or less likely, depending on the size of the potential partner's military contribution. Finally, the promise of successfully building a coalition makes states more willing to initiate or "select into" crises than they would be otherwise. As noted in the appendix, this has implications for the process that generates the sample of crises used to analyze this chapter's empirical model, requiring the estimation of a selection model (also in the appendix) as a robustness check.

Coalition Formation in Crises

Proposition 3.1 summarizes each player's strategy in the unique Subgame Perfect Equilibrium, which requires that all strategies be sequentially rational – that is, that no player can commit ex ante to follow through with an action that it would not be willing to carry out if given the opportunity. As such, an SPE outlines the conditions under which (a) L initiates a crisis or tolerates the status quo, (b) L proposes a coalition or acts unilaterally, and (c) P accepts or rejects L's coalition proposal.

Proposition 3.1. *The following strategies constitute the game's unique Subgame Perfect Equilibrium. When $\beta \leq \beta^{\dagger}$, L initiates a crisis iff $q \leq q_L$ and acts unilaterally. When $\beta > \beta^{\dagger}$, L initiates a crisis iff $q \leq q_{LP}$ and proposes $s^* = \hat{s}$. P accepts any proposal $s \leq \hat{s}$ and rejects otherwise.*

When will L and P form a coalition? Proposition 3.1 states that they do so when some side payment proposal exists that P is willing to accept and that L is willing to make. Formally, this occurs when their preferences are sufficiently similar, or when

$$\beta > \frac{c_P - (p_{LP} - p_L)}{\gamma(p_{LP} - p_L)} \equiv \beta^{\dagger}. \tag{3.3}$$

The reasoning behind this result is straightforward: when the lead state and potential partner have similar foreign policy preferences, P requires less compensation to join the coalition, and the lead state is happier to take on a partner when doing so does not require large side payments or substantial compromises over the management of the crisis. To see how the diversity of preferences shapes coalition formation, note that L's equilibrium level of proposed compensation leaves it with

$$s^* = \frac{p_{LP}\left(1 - \gamma\left(1 - \beta\right)\right) - \beta\gamma p_L - c_P}{p_{LP}\left(1 - \gamma\right)},$$

which increases in β. Thus, L can secure P's cooperation at ever lower prices the more their preferences align over the outcome of the crisis. All else being equal, then, states building coalitions are more likely to take on partners who share their preferences – as opposed to states with more divergent interests that would require more substantial side payments or policy compromises in order to ensure their cooperation.

Proposition 3.2. *The more P's foreign policy preferences align with L's, the more willing is L to propose and secure P's acceptance of a coalition.*

This result is generally unsurprising; why, after all, would states build coalitions with partners that desire wildly different outcomes of the crisis? However, a closer look at Inequality (3.3) shows that both the lead state's and partner's military capabilities also play a role in coalition formation, to the point that even similar preferences may sometimes be insufficient to ensure that L and P reach an agreement. Note that the military benefits of taking on a partner, defined by the difference $(p_{LP} - p_L)$, enter both numerator and denominator, although the effect in each case is to render the coalition formation constraint easier to satisfy as those military benefits increase. First, consider L's own military capabilities, m_L. The difference between p_{LP} and p_L, both of which depend on m_L, decreases as L becomes stronger, which reduces the return on any given side payment; this is clearest in the numerator, where a decrease in $(p_{LP} - p_L)$ makes it harder to cover the costs of side payments that increase directly in P's costs for participation, c_P. Thus, conditional on entering a crisis, *militarily stronger states are less willing to form coalitions than militarily weaker states* are, all else being equal. However, as discussed later, the link between L's military strength and initiating crises means that this relationship may be difficult to uncover empirically.

Proposition 3.3. *The greater L's own military capabilities m_L, the higher is $\beta^{\dagger}$, rendering L both (a) less willing to propose and secure P's acceptance of a coalition and (b) less willing to tolerate a partner with preferences increasingly divergent from its own.*

This also implies that, even when partners with similar preferences are available, powerful lead states can afford to be selective, as they are freer to forego coalition-building when required compensation outstrips meager military contributions. The United States did just this when it turned down a number of potential military contributions during the Korean War, including Pakistan's offer of ground troops, which came with the price tag of a formal defense commitment and support against India in Kashmir (Stueck 1995, pp. 73–74).

Of course, after Chinese intervention later in the war, the United States would come to regret turning down as many offers of assistance as it did in the fall of 1950, but at the time – just as United Nations forces turned the tide and began pursuing North Korean forces above the 38th parallel – very favorable military prospects militated against substantial concessions in return for additional cooperation (*ibid.*, ch. 5). Similarly, while the Central Powers were able to bring Turkey into their coalition in the First World War with promised concessions in the Balkans, as well as cash and military aid (Hastings 2013, p. 414), the Allies hoped to bring Japan fully into the war but could not find a sufficient "bribe," or side payment, to ensure its cooperation beyond occupying German colonial possessions in East Asia (p. 543). Finally, even Nazi Germany made hard calculations about the military value of potential coalition partners and required compensation during the Second World War, as "[b]y February 1941 . . . [Hitler's] interest in Gibraltar waned, and with it Germany's willingness to pay an extravagant price for Spanish belligerency" (Hastings 2012, p. 111).

Next, consider P's potential contribution to the lead state's chances of success in the crisis, m_P, which straightforwardly increases p_{LP}. As m_P grows, so does the difference in L's chances of success across multilateral and unilateral action, $(p_{LP} - p_L)$; thus, the coalition formation constraint falls, becoming easier to satisfy with ever lower values of β – that is, with ever more divergent preferences between lead state and partner. In other words, the more a state benefits from a partner's military assistance, the less sensitive it becomes with respect to the cost of securing that partner's cooperation, taking on partners with preferences increasingly divergent from its own. This means (a) that lead states prefer powerful partners over weaker ones, all else being equal; *and* (b) that a partner's potential military contribution should mediate the effect of preference divergence.

Proposition 3.4. *The greater P's military capabilities m_P, the lower is $\beta^\dagger$, such that L is both (a) more willing to propose and secure P's acceptance of a coalition and (b) more willing to tolerate a partner with preferences increasingly divergent from its own.*

As stated formally in Proposition 3.4, the threshold over β above which L proposes a coalition, or $\beta^\dagger$, decreases as P's military capabilities increase. In other words, the more influential is P's contribution to L's chances of success in the crisis, the less concerned L becomes about the cost of compensation, taking on partners with preferences increasingly divergent from its own. While the lead state must yield a larger share of the pie generated by the crisis, the increased costs of cooperation are easily covered by the fact that the pie is made larger in expectation due to P's military contribution. This suggests that a potential partner's military power affects both the probability of coalition formation and the observed diversity of preferences within those coalitions that form;

would-be coalition leaders prefer to minimize the compromises they make to enhance their military power, but as they grow more desperate militarily, or as potential partners offer ever larger or decisive military benefits, they become more willing to take on partners with divergent preferences, making ever larger compromises in order to do so.

To see how the logic of coalition formation works in practice, consider Saudi Arabia's participation in the 1991 Gulf War. In the run-up to war, the United States knew that the continued support of the Saudi royal family was critical; only Saudi Arabia bordered both Kuwait and Iraq to the south, allowing it to provide essential basing, staging, and logistical support for what would ultimately be a massive military effort. Despite the threat to their oil fields posed by an Iraqi army sitting astride Kuwait in late 1990, which made cooperating with the United States attractive, the Saudis were also concerned about the consequences of a costly and chaotic breakup of Iraq that might follow from toppling the Ba'athist government in Baghdad. Keenly aware that Saudi support was the sine qua non of an effective liberation of Kuwait, the United States went to great pains to compensate the Saudis for their cooperation, limiting its demands to the restoration of the Kuwaiti royal family and the degradation of the Iraqi Army – at the time considered battle-hardened after nearly a decade of war with Iran and touted as the fourth largest in the world (Bush and Scowcroft 1998, pp. 313, 491).[13] In the absence of Saudi cooperation, liberating Kuwait would have required either (a) an amphibious assault from the Persian Gulf, which would have proven much more costly – and riskier – than an overland invasion; or, even less attractive, (b) a cross-country assault launched from Turkey. Both alternatives were clearly less preferable than accepting limited war aims in return for the opportunity to launch the attack from Saudi territory. Therefore, given Saudi Arabia's uniquely valuable military contribution, one that no other state could provide, the United States compromised on its war aims in order to secure military cooperation.

Next, consider the military strength of the very state against which L might seek assistance: the target. Ostensibly, we might expect that stronger targets would foster a greater willingness to propose generous terms that secure P's cooperation, but close examination of how $\beta^\dagger$ moves with respect to m_T shows that this logic is faulty. Specifically, the probability of coalition formation can increase or decrease in the target's military capabilities, depending on the level of those same capabilities (m_T).

Proposition 3.5. *The effect of T's military capabilities depends on m_T itself. When $m_T < \sqrt{m_L (m_L + m_P)}$, $\beta^\dagger$ decreases in T's military capabilities, rendering L more likely to form a coalition, and when $m_T > \sqrt{m_L (m_L + m_P)}$, $\beta^\dagger$ increases in T's military capabilities, rendering L less likely to form a coalition.*

---

[13] See also Atkinson (1993, pp. 298, 299) and Kreps (2011, p. 27).

Proposition 3.5 states that the relationship between target strength and coalition formation is nonlinear; specifically, coalition formation should be easiest when T's military capabilities are neither too great nor too small. When the target is weak, L has little need of coalition partners, and it finds acting as a singleton relatively attractive; why compensate a partner that one simply may not need? On the other hand, when T is quite strong, partners may be difficult to come by, precisely because they will be reluctant to join a coalition that is unlikely to succeed against its opponent. When m_T falls in a middling range, however, a potential partner can have more of a material effect on the outcome of the crisis *and* L is more willing to pay for its cooperation. Therefore, the model predicts that there should be no consistent linear relationship between target strength and coalition formation, but a nonlinear one. However, as is the case with predictions over the effect of the lead state's strength derived in the next section, the effect of target strength on crisis initiation may pose problems for uncovering its effect on coalition formation in observational data.

Finally, Inequality (3.3) shows that the relative size of the public component (γ) of the stakes also has an effect on the formation of coalitions. Generally, we might expect that a large public component – that is, some issue between L and T that is especially salient for P – should make cooperation of any kind more difficult (Olson 1965); L's erstwhile partner stands a chance of enjoying the substance of the policy outcome even if it refuses to join the coalition and thereby saves the costs of participating in the coalitional effort. Yet this line of argument ignores the lead state's ability to shape the terms of P's cooperation, which it may do by offering concessions or compromises that take the form of a "selective incentive" that can outweigh the individual temptation to refuse cooperation (*ibid.*, p. 51). In fact, there also exist conditions under which an increasing public component can make coalitions *easier* to form, while under others it discourages cooperation, as a more conventional account would lead us to expect. As stated in Proposition 3.6, the effect of the relative size of public and private components depends on the extent of the military boost provided by P's cooperation and P's own costs for participation, which shape the side payment required to secure its cooperation.

Proposition 3.6. *The effect of the public element γ of the stakes depends on $p_{LP} - p_L$. When $p_{LP} - p_L \geq c_P$, $\beta^\dagger$ weakly increases in γ, rendering L less willing to form a coalition, but when $p_{LP} - p_L < c_P$, $\beta^\dagger$ decreases in γ, rendering L more willing to form a coalition.*

Beginning with the case in which P's military contribution to the coalition would be large, or $p_{LP} - p_L \geq c_P$, an increasing public component makes coalition formation difficult, because P can both demand a large premium for its cooperation *and* credibly refuse to participate; given such a sizable public goods component of the stakes, it enjoys the outcome whether or not it joins the coalition. By contrast, a relatively smaller public share implies a larger private share, which undermines the partner's threat to remain on the sidelines. In this case of a relatively large military contribution, then, coalitions can only be built around sizable policy concessions, which ensures that few are likely to form in the first place. However, when P makes a smaller military contribution, or $p_{LP} - p_L < c_P$, the story is different. The larger the public share, the more L is willing to concede larger shares of the smaller private component in order to entice P to cooperate, overcoming the partner's otherwise strong incentives to stay on the sidelines, because it requires a much smaller premium in return for its participation. Coalitions should thus be more likely as the relative share of public stakes rises, because they can be based on relatively inexpensive side payments. On the other hand, when the private stakes of the crisis are larger and P's military contribution is not too large, then L is less likely to find side payments cheap enough to justify bringing P on board.

Therefore, the nature of the international stakes and the size of the partner's military contribution shape both (a) the side payments P can demand and (b) L's willingness to make them, suggesting an interactive effect on the formation of coalitions: an increasing public goods component makes weaker partners more attractive and stronger partners less attractive. For example, interventions aimed at producing public goods or eliminating public bads – for example, stopping a genocide or stemming refugee flows from a civil war – might find lead states taking on weaker partners than for missions aimed at conquest or regime change.

What does this mean for patterns of coalition-building and international military cooperation? Most broadly, and as discussed further in Chapter 6, it suggests that scholarship on military multilateralism might benefit from a focus on the types of issues over which states contend. While distinctions between territorial and nonterritorial disputes (Senese and Vasquez 2008, Vasquez 2009) and between security and economic issues (Keohane 1984, Keohane and Nye 1977) have shed light on the chances of war and the possibility of cooperation, respectively, the relative share of public and private elements in a given dispute may illuminate these processes further by helping explain when and why coalitions form in some crises and not in others. For example, holding the supply of potential partners constant, Proposition 3.6 suggests that states will seek out relatively weaker partners when they pursue goods with a large public component, because weaker partners will drive easier bargains over smaller private goods components; however, perhaps counterintuitively, when the share of private goods is larger, states will seek out stronger partners, because their threats to refuse cooperation are less credible, allowing the lead state to drive a harder bargain, *and* because concessions out of a larger private pie are relatively less painful.

Coalitions and Crisis Initiation

Moving back up the game tree in Figure 3.1, we can now consider the lead state's initial choice over entering the crisis or tolerating the status quo. How does the anticipation of coalitional or unilateral action affect L's decision to initiate a crisis in the first place? In other words, when will states initiate crises and go on to build coalitions? This question is interesting substantively, as it sheds new light on the general relationship between military cooperation and international conflict, but it is also important to consider analytically, to the extent that it can inform our understanding of the selection process by which states come to enter the observed sample of crises analyzed in the following section. Analysis of this initial decision shows that the availability of a cheap coalition partner does, indeed, encourage states to select themselves into crises. Further, it also shows that L's military capabilities encourage crisis initiation, even as they go on to discourage coalition formation. This has serious implications for the expected relationship between capabilities and coalition formation in samples of states already participating in crises.

Proposition 3.1 states that L initiates a crisis when it is dissatisfied with the international distribution of benefits – that is, when its payoff for entering a crisis and engaging in military coercion is greater than its payoff for tolerating the status quo.[14] However, whether it finds the status quo tolerable depends on whether it expects to build a coalition. When L and P will not go on to form a coalition, the lead state initiates a crisis when its expected utility for the crisis game as a singleton yields more than the status quo, or simply when $q < p_L - c_L \equiv q_L$. However, when L does expect to secure P's military cooperation and build a coalition, the lead state initiates a crisis when

$$q < p_{LP} - c_L - c_P + \beta\gamma(p_{LP} - p_L) \equiv q_{LP}, \tag{3.4}$$

or, again, when initiating crisis has a higher expected value than the status quo.

Whether a coalitional crisis is worth initiating, however, depends on some particular characteristics of the potential partner. As indicated by Inequality (3.4), the threshold below which the lead state is willing to initiate a crisis decreases in the partner's costs for participation, c_P, but increases in preference similarity, β. Thus, when partners are available on the cheap in terms of compensating them for their cooperation, potential coalition-builders are comparatively more willing to initiate crises. On the other hand, when available partners require larger political concessions, initiating crises is far less attractive, and L tolerates the status quo rather than mortgage whatever it might gain during the crisis to ensure a partner's cooperation.

The initiation constraint also becomes easier to satisfy as P's military contribution grows larger; the more powerful an anticipated partner, the more willing the lead state is to initiate a crisis. However, as stated in Proposition 3.7, this effect diminishes as the lead state's own military power increases. Weaker states are in general unlikely to find crisis initiation attractive, but their willingness to do so can be substantially increased by the availability of a

---

[14] Note that this differs slightly from Powell's (1999) definition of dissatisfaction, which requires that a state prefers war to the status quo.

partner. For powerful states, however, the same promise of military cooperation is less meaningful, as they are already likely to find selecting into the crisis
attractive in the first place.

Proposition 3.7. *The greater P's military capabilities m_P, the higher is q_{LP},
rendering L more willing to initiate a crisis. However, this effect diminishes as
L's own capabilities m_L increase.*

Next, some features of the principal dyad, L and T, also figure prominently
in the decision to initiate a crisis. In each case, the effect of military capabilities
when L will act as a singleton is straightforward, but the prospect of building
a coalition complicates matters substantially.

Proposition 3.8. *The greater L's military capabilities m_L, the higher is q_L,
reindering L more willing to initiate a crisis. However, there is no consistent
bivariate relationship between m_L and q_{LP}.*

Proposition 3.9. *The greater T's military capabilities, the lower is q_L, rendering
L less willing to initiate a crisis. However, there is no consistent bivariate
relationship between m_T and q_{LP}.*

Proposition 3.8 states that, when L expects to act alone, increasing military
capabilities make it more willing to initiate a crisis. Likewise, Proposition 3.9
indicates that L is generally hesitant to initiate crises against militarily stronger
targets if it expects to act as a singleton. However, when L expects to take on a
partner, it may be more or less willing to initiate a crisis as a function of dyadic
military capabilities, depending on the values of a complex combination of its
own military capabilities, the target's, the partner's, and the specific price of
cooperation.

While this makes simple predictions difficult to derive, it has other subtle
but significant empirical implications, because (a) increasing lead state military
capabilities also discourage coalition building once inside a crisis and (b) the
effect of target capabilities is already predicted to be nonlinear. This makes
drawing inferences about the effect of L's and T's capabilities on coalition
formation difficult in a sample of observed crises, because it is difficult to
assess precisely how states select into crises, at what rates, and what this process
means for the observed distributions of military strength variables in a sample
of crises. I discuss these challenges at greater length later.

Finally, comparing the two constraints over crisis initiation shows that L is
generally more willing to initiate a crisis when it expects to build a coalition,
or when

$$p_{LP} - c_L - c_P + \beta\gamma(p_{LP} - p_L) > p_L - c_L.$$

Algebra shows that this inequality holds as long as $\beta > \beta^\dagger$, which is precisely the
condition supporting coalition-building given in Proposition 3.1. As a result,
Proposition 3.10, which indicates that states are more likely to initiate crises

when they expect military cooperation than they they expect to act unilaterally, follows directly.

Proposition 3.10. *States are more willing to select themselves into crises when they expect coalitional support than when they expect to act unilaterally.*

Thus, in addition to whatever political benefits that states derive from fighting with "friends and allies," their availability as coalition partners should make states *more* willing to initiate crises than they would be if they knew they would have to act alone.[15] In this sense, the very possibility of cooperation is associated with an increase in the chances of conflict, at least at the level of the willingness to enter or initiate crises. However, as shown in Chapter 4, this need not mean that coalition partners have a similar impact on the escalation of these crises to full-scale war.

Lastly, Propositions 3.7–3.10 imply that problems of sample selection might plague any empirical assessments of coalition formation that begin with a sample of international crises. Specifically, if states are more likely to initiate crises when they also expect to form coalitions *and* if errors in prediction across one stage are correlated with errors in the other, then inferences over the effects of some independent variables – for example, military capabilities – might be suspect (see Angrist and Pischke 2009, Heckman 1979).[16] Further, the effects of L's and T's military capabilities in a sample of crises are ambiguous, because m_L has opposing effects on crisis initiation and coalition formation, while the effects of m_T are highly nonlinear. Fortunately, these propositions do indicate the proper steps to take in order to assess potential selection bias in the sample of crises analyzed later: estimation of a selection model that captures the extent to which unobservable factors, or variables omitted from the model, lead to correlated errors across the two stages. Therefore, I estimate a probit model with sample selection in the appendix, using the data described in the following section, to show that there is no systematic relationship in the errors for crisis onset and coalition formation (controlling for military capabilities), which increases confidence in inferences drawn over coalition formation in the remainder of this chapter – as well as models of escalation and expansion in subsequent chapters.

Summary of Empirical Implications

To summarize the implications of the theory, states would like to form coalitions with partners that share their foreign policy preferences, as this limits the

---

[15] See Chapman and Wolford (2010) for an argument that international organizations like the UNSC can also encourage conflict when their support can lower the costs of war for a strategic state intent on winning its approval.

[16] It is possible, of course, that sample selection works in the opposite direction: if states or their leaders have reputations for managing coalitions effectively (Huth 1997, p. 76), then lead states might be hesitant to initiate crises against them in the first place, further complicating the relationship between coalition-building and crisis initiation.

extent of the compromises required to ensure military cooperation, but they also tend to pursue strong partners as well. As a result, lead states become less selective with respect to their prospective partners' preferences the more they can aid the military effort, suggesting that powerful potential partners are both more likely to be secured and that they will be associated with greater levels of preference diversity in observed coalitions. Additionally, the size of the partner's military contribution determines whether an increasing public-goods component of the international stakes makes coalition formation more or less likely. Finally, states will also be more likely to initiate crises when they expect to be able to build coalitions than when they expect to act alone. I turn in the next section to an empirical evaluation of the first three implications.

3.3 EMPIRICAL MODELS OF COALITION FORMATION

When do states build coalitions to improve their chances of success in international crises, and what accounts for the partners they choose? Extant empirical models posit that the issues at stake (Tago 2007), operational needs and threat levels (Kreps 2011), domestic politics (Mousseau 1997, Vucetic 2011), prior commitments (Corbetta and Dixon 2004, Pilster 2011), and the approval of international institutions (Chapman 2011, ch. 5) all play a role, but the theory presented in this chapter suggests that there are specific characteristics of potential partners that affect (a) a lead state's desire to gain their cooperation and (b) the ease with which their cooperation can be purchased, both of which have implications for the probability with which coalitions form and the diversity of preferences among their members. The theory predicts that lead states in a crisis will be more likely to form a coalition with a particular partner when that partner's preferences are similar to its own, although the coalition-builder will tolerate increasingly divergent preferences in militarily powerful partners – which are attractive partners in their own right. In this section, I use the data described in Chapter 2, as well as data on state military capabilities (Singer, Bremer, and Stuckey 1972) and revealed foreign policy preferences (Strezhnev and Voeten 2013), to test these hypotheses.

To explore coalition formation, I sample on crises, which is less problematic for hypotheses over preference diversity and partner characteristics than it is for those related to leader and target characteristics. Thus, in the following section, I state only hypotheses related to the former two variables, and I follow the same logic in choosing the hypothesis to test about preference diversity in coalitions that do form. However, throughout the analysis I leverage insights gained from a selection model analyzed in this chapter's appendix, which takes a different unit of analysis (leader-target directed dyads) to estimate potential correlations in the errors across the onset of a crisis and whether L forms a coalition with *any* partner. I find no such correlation, which increases confidence in estimates generated from the crisis-only sample, but I am also cautious to draw inferences over leader and partner capabilities given that the selection model and the models in this section use a different unit of analysis.

3.3.1 Hypotheses

The theoretical model generates empirical implications over two different outcome variables: coalition formation between L and P on the one hand, and the diversity of preferences inside coalitions on the other. Before discussing the research designs appropriate for each empirical model, I use this section to translate the logic of the theoretical model into hypotheses that can be tested on the data described in the previous chapter. Those variables that the theory indicates are most prone to selection problems – particularly L's and T's military capabilities – are not stated as formal hypotheses, but I do discuss what can be learned from the inclusion of those variables where relevant.

Beginning with coalition formation, I test three hypotheses, based respectively on a potential partner's military capabilities and the difference between L's and P's foreign policy preferences. As discussed earlier, increases in the former make L more willing to compromise in order to secure cooperation, while increases in the latter make L less willing to propose a coalition to P. However, lead states should become less concerned about preference divergence for more powerful potential partners, implying that these two factors have a conditioning, or interactive, effect as well.

Hypothesis 3.1. *During a crisis, the probability that L forms a coalition with P increases in P's military capabilities.*

Hypothesis 3.2. *During a crisis, the probability that L forms a coalition with P decreases in the divergence of foreign policy preferences between L and P.*

Hypothesis 3.3. *During a crisis, the effect of preference diversity on the probability of coalition formation diminishes in P's military capabilities.*

Recall that the theoretical model makes divergent predictions over the effect of the lead state's military capabilities on coalition formation and crisis initiation, which means that I can derive no firm prediction over its effect in a sample of crisis participants. As such, I do not assess predictions over the probability of coalition formation as a function of L's military power, although I do discuss its effects at greater length later. I also include T's capabilities as a control variable, but the selection process suggests that any related inferences might be as problematic as those based on L's power.

Second, specific lead state and partner characteristics should also be associated, on average, with more or less diversity of preferences among observed coalition members. Therefore, among states that join coalitions, any factor that increases their military value should be associated, on average, with an increase in the difference between that state's foreign policy preferences with the lead state's. Likewise, any factor that limits the need for military assistance should decrease the observed differences between lead state and partner preferences. Thus, Hypotheses 3.4 follows directly from the logic of Proposition 3.3.

Hypothesis 3.4. *Among observed coalitions, the difference between L's and P's foreign policy preferences increases in P's military capabilities.*

Put differently, a state's selectivity when it comes to choosing partners – its willingness to pay a certain price for cooperation – will vary with just how much it values the services of any given partner: the more powerful the potential partner, the more its preferences should diverge, on average, from the lead state that induced it to join the coalition. While Proposition 3.4 also indicates that L's own capabilities should produce a complementary relationship, decreasing the diversity of preferences in observed coalitions, I do not specify an additional hypothesis because of the potential selection issues pointed to by Proposition 3.8. However, I still include it as a control in the empirical model. I turn in the following section to a discussion of how I design empirical models to test each of these hypotheses.

3.3.2 Research Design

Testing Hypotheses 3.1–3.4 requires data on coalition participation in crises, as well as the identity, military capabilities, and foreign policy preferences of potential coalition members. In this section, I discuss the construction of the dataset, theoretical variables, control variables chosen to reduce the chances of finding spurious relationships, and the specification of the two empirical models used to test each set of hypotheses.

To create the necessary sample, I begin with the list of coalitions and their participants discussed in Chapter 2, which will be used to create the dependent variable. I then reference the principal dyad in the crisis – that is, the two states on opposite sides of the central issue – who are the most plausible candidates to build coalitions as lead states. The coding rules and list of principal belligerents are discussed in Chapter 2, but to cite an example of how the rules work in practice, consider ICB crisis #342, in which a coalition of Chad and France confronted Libya with a threat of war over the latter's cross-border support for Ogaden insurgents. In this crisis, Chad and Libya are the principal belligerents, and each of the two is observed once as a potential coalition builder L and once as a target T. Thus, when Chad is observed as L, it is observed as gaining a coalition partner in France (P), while Libya builds no coalition when observed as L.

The next step is to identify for each principal disputant a set of potential coalition partners from which it may draw. This implies a dataset of directed-dyadic lead state-partner observations, two sets for each crisis, where each potential coalition-builder is paired with a given number of potential partners. To create this sample, I identify three criteria, any one of which is sufficient to mark a state as a potential partner for *L*. The criteria are: (a) any state that borders *L* or *T* by land or fewer than 400 miles of water, according to the Correlates of War project's contiguity data (Stinnett et al. 2002); (b) any state within *L* or *T*'s geographic region, based on the Correlates of War project's coding scheme (Sarkees and Wayman 2010); and (c) any major power (COW 2011, Gleditsch and Ward 2001).[17] This is slightly more permissive than Chapman's (2011) definition, which includes only those states bordering the target and major powers. My selection criteria strike a balance between considering too few potential partners on the one hand and risking aggregation bias on the other by including too many implausible partners. Notably, this sampling scheme captures all but four observed coalition members in the data, without having to select the sample on measures of foreign policy affinity – say, alliances – which would be inappropriate given the role played by preference divergence in the theory.[18] To return to crisis #342, principal belligerents Chad and Libya have 64 and 67 potential partners each, drawn from their immediate geographical neighborhood, regions, and contemporary major powers, resulting in 131 total dyads of lead state and potential partner associated with this particular crisis.

The outcome variable is $\text{Join}_{L,P}$, which indicates the formation of a coalition as defined in Chapter 2. Specifically, $\text{Join}_{L,P} = 1$ when *L* and a potential partner in a dyad form a coalition, and $\text{Join}_{L,P} = 0$ otherwise. Returning again to crisis #342, Libya scores a zero in every Libya-partner dyadic observation, since none of its potential partners form a coalition with it. On the other hand, the Chad-France dyad is coded as $\text{Join}_{L,P} = 1$, since France joins on Chad's side, while all other Chad-partner dyads receive a zero. Thus, when $\text{Join}_{L,P} = 1$, the pair of states in question achieves a mutually favorable exchange of military cooperation and political compromise such that they form a coalition, with the understanding that they will both apply military pressure to the opposing side if the target resists their collective demands.

To clarify how the data are structured around leader-partner dyads and the coalitions that form among them, Table 3.1 presents a truncated example of how the data are structured around ICB crisis #183, touched off in 1961 with Iraq's threat to invade Kuwait following the latter's announcement of independence at the end of the British protectorate. Note that the table uses only the neighbor and great power criteria to keep the size manageable. Kuwait and Iraq are the principal belligerents, identified once each as *L* and *T* and

---

[17] I adjust regions slightly, such that Western and Eastern Europe are considered part of the same region, as well as East Asia and Oceania. Otherwise, regions are North America, South America, the Middle East/North Africa, and the remainder of Africa. The shared region criterion captures a surprising number of coalition members that the pure distance measures cannot, such as the Western European states joining the coalition arrayed against Serbia in the Kosovo War.

[18] Those partners not captured by the scheme are (a) Cuba joining Angola in ICB #260, (b) Mauritania joining Morocco in ICB #261, (c) Belgium joining Zaire in ICB #292, and (d) Libya joining Uganda in ICB #296.

TABLE 3.1. *The Structure of the Coalition Formation Dataset, Using Only Neighbor and Great Power Partner Criteria*

Crisis #	Leader (*L*)	Target (*T*)	Partner (*P*)	Join$_{L,P}$
183	Kuwait	Iraq	Saudi Arabia	0
183	Kuwait	Iraq	China	0
183	Kuwait	Iraq	Jordan	0
183	Kuwait	Iraq	France	0
183	Kuwait	Iraq	Japan	0
183	Kuwait	Iraq	Soviet Union	0
183	Kuwait	Iraq	Egypt	0
183	Kuwait	Iraq	Turkey	0
183	Kuwait	Iraq	Iran	0
183	Kuwait	Iraq	United States	0
183	Kuwait	Iraq	Syria	0
183	Kuwait	Iraq	United Kingdom	1
183	Iraq	Kuwait	Saudi Arabia	0
183	Iraq	Kuwait	Egypt	0
183	Iraq	Kuwait	Iran	0
183	Iraq	Kuwait	United Kingdom	0
183	Iraq	Kuwait	Soviet Union	0
183	Iraq	Kuwait	France	0
183	Iraq	Kuwait	China	0
183	Iraq	Kuwait	Syria	0
183	Iraq	Kuwait	Japan	0
183	Iraq	Kuwait	Jordan	0
183	Iraq	Kuwait	Turkey	0
183	Iraq	Kuwait	United States	0

observed as many times as there are potential partners (by these limited criteria, twelve each). For each pairing of lead state and potential partner, the outcome variable Join$_{L,P}$ takes on a value of 1 when the partner joins a coalition with *L* in the crisis under observation and zero otherwise, which shows that the United Kingdom joined a coalition with Kuwait, while Iraq had no coalition partners in the crisis.

Next, the theory implies three main theoretical variables: (a) potential partners' military capabilities, m_P; (b) the extent of preference divergence between *L* and *P*; and (c) an interaction between the two. In this and all subsequent empirical models, I use the Correlates of War Project's Composite Index of National Military Capabilities (Singer 1987, Singer, Bremer, and Stuckey 1972) as a measure of state military capabilities, labeled CINC$_P$ for partners. The same term is subscripted with *L* and *T* for leaders and targets where relevant. The CINC index gives a state's share of the total military capabilities in the international system in a given year, where capabilities are a function of total national population, urban population, iron and steel production, energy consumption, military personnel, and military spending; in other words, it is designed to measure a state's ability to mobilize for and fight an interstate war – the ultimate outcome whose shadow looms over processes of coalition-building and crisis bargaining.

To measure revealed foreign policy preferences, I use ideal points estimated from United Nations General Assembly (UNGA) voting patterns in the years 1946–2001. I use two sources, Strezhnev and Voeten (2013) and Reed et al. (2008), although due to its greater temporal coverage, I present results based on the former's estimates. Strezhnev and Voeten's (2013) UNGA ideal points are estimated along a single dimension, one that captures degrees of support for or opposition to the United States–led liberal order in which the UN originated and that it continues to represent.[19] Given the wide range of issues voted on, as well as "incentives to vote in self-interested ways even if the issue at hand does not bear on a particular state," these votes are an effective indicator of revealed foreign policy preferences (Reed et al. 2008, p. 1207). The logic behind the measure is thus quite simple: the more often states agree on the wide variety of issues brought before the UNGA, the more likely they are to agree on any given issue, the current crisis under observation in particular. To generate the variable Divergence$_{L,P}$ for each leader–potential partner dyad, I take the absolute value of the difference between L and P's ideal points, which produces a rough measure of the distance between their evaluations of the international distribution of benefits. The smaller the value of Divergence$_{L,P}$, the more similar are preferences in the dyad; the larger the value, the more their preferences diverge, and the more likely they are to disagree over the issues under dispute. I also lag these values by a year to avoid conflating votes on the issues involved in the crisis with actions taken within the crisis, such as coalition formation.

Finally, I include a set of control variables that are plausibly linked to both preference divergence and coalition formation. First are the lead state's own and the target's military capabilities, CINC$_L$ and CINC$_T$, which influence a lead state's chances of success and, thus, the desirability of a partner. Allies$_{L,P}$ indicates whether L and P share either a defense pact or an entente according to the Correlates of War's alliance data (Gibler and Sarkees 2004), as allies might find working together more efficient than nonallies (Morrow 1994, 2000) and tend to share foreign policy preferences (Gibler and Rider 2004). Next, Chapman (2011) has shown that potential partners are more likely to join the sides of those states that win approval from a conservative UN Security Council, so I include the dummy UNSC Support$_L$ that indicates whether the Security

---

[19] The Reed et al. (2008) estimates are also single-dimensional, dominated by the Cold War rivalry between the United States and the Soviet Union before 1990 and the American-Russian rivalry afterwards. Prima facie, this dimension is very similar to Strezhnev and Voeten's (2013), and in fact the two ideal point variables are highly correlated (roughly −0.95).

Council either condemns the actions of the target or authorizes L's use of force according to ICB crisis summaries (see also Wolford 2014a). UNSC Support$_L =$ 1 if the UNSC passes a resolution that (a) explicitly condemns the actions of side 2 in starting the crisis, (b) calls for only side 2 to withdraw forces or cease hostilities, (c) imposes or escalates a sanctions regime against side 2, or (d) authorizes side 1's use of military force. If the UNSC is silent or if it condemns or calls for action from both sides, as it did in the 1969 Egyptian-Israeli War of Attrition (ICB #232), this variable equals 0. Other controls indicate whether leader, partner, and target are democracies according to the Polity-IV combined democracy score (Marshall and Jaggers 2009), whether leader and target and whether partner and target share a land border (Stinnett et al. 2002), an indicator of whether the United States is observed as L, to ensure that the United States' ostensibly unique levels of coalitional activity (Kreps 2011, Tago 2007) do not contaminate the results, and a marker for the post–Cold War era, when diplomatic multilateralism became more common with the end of the Soviet veto in the Security Council (Voeten 2005).

With sample and variables defined, we can state the full statistical specification for the model of coalition formation,

$$\Pr(\text{Join}_{L,P}) = \Phi(\alpha + \beta_1 \text{Divergence}_{L,P} + \beta_2 \text{CINC}_P +$$
$$\beta_3 \left(\text{Divergence}_{L,P} \times \text{CINC}_P\right) + \beta \mathbf{X}_{si} + \varepsilon_i), \tag{3.5}$$

where Φ is the CDF of the standard normal distribution, implying a probit model, and $\mathbf{X}_{si}$ is a vector of control variables measured by side s and crisis i. I also estimate Huber-White robust standard errors to account for intergroup correlations and cluster them by individual crisis ε_i to account for the double-counting of each principal belligerent as a potential coalition leader.[20]

The empirical model of preference diversity inside coalitions takes the absolute value of the difference between L and P's UNGA ideal points, Divergence$_{L,P}$, if the two form a coalition – that is, if $\Pr(\text{Join}_{L,P}) = 1$ – as the outcome variable. The main theoretical variable is the partner's military capabilities (CINC_P); control variables include lead state and target capabilities, as well as indicators for contiguity between partner and target, an alliance between L and P, whether L is the United States, whether it receives UNSC support, and whether the crisis occurs after the Cold War. Thus, the empirical specification is

$$\text{Divergence}_{L,P} = \alpha + \beta_1 \text{CINC}_P + \beta \mathbf{X}_{si} + \varepsilon_i \tag{3.6}$$

---

[20] I also estimate alternative models where standard errors are clustered on the lead state to account for any persistent state-specific tastes for coalitions or unilateral action, as well as a random-effects version that estimates a separate random intercept for each crisis, and the results are substantively similar.

implying a simple ordinary least squares (OLS) regression, since the differences between ideal points are theoretically continuous.[21] As before, X_{si} is a vector of control variables, and I cluster Huber-White robust standard errors by the individual crisis ε_i.

Below, I first estimate the probit model of coalition formation to test Hypotheses 3.1–3.3, then estimate the OLS model of coalitional preference diversity to test Hypothesis 3.4.

3.3.3 Results: Coalition Formation

Table 3.2 presents the results of the probit model of coalition formation. Generally, the coefficients on the control variables behave as expected. First, among domestic political factors, only the partner's regime type has a statistically discernible effect ($p < 0.1$) on the probability of coalition formation: democratic partners appear more eager than nondemocratic countries to join coalitions. Therefore, even as democratic states are not uniquely likely to join wartime coalitions with one another (Reiter and Stam 2002, ch. 4), they are uniquely likely to join coalitions in general at the stage of crisis bargaining. Next, explicit support from the UN Security Council increases the probability that a coalition forms, consistent with the international audience variant of Chapman's (2011) theory, although its effect, like that of Democracy$_P$ is statistically discernible only at the $p < 0.1$ level. Given the rarity with which the Security Council weighs in on interstate crises, even limited statistical significance is impressive; however, Chapman and Wolford (2010) argue that states are only likely to seek approval when they expect a favorable vote, which might overstate the observed effect of UNSC approval. Finally, the United States appears no more likely than other states to build coalitions; to the extent that the United States builds coalitions often, it appears to be the result of other factors already accounted for in the empirical model.[22]

Two other coefficients stand out among the controls. First, the target's neighbors are uniquely likely to join coalitions with the lead state, as indicated by the positive and statistically significant coefficient ($p < 0.01$) on Contiguity$_{P,T}$. While this could be indicative of L's desire to gain military partners that afford easy access to their targets – as the United States did with Kuwait in 2003 – it may also reflect the fact that partners are more easily compensated when they oppose their neighbors as much as L does, so it is possible that this coefficient is consistent with either Hypothesis 3.1 or 3.2; the model offers no way to adjudicate between the two. Second, the presence of an alliance between L

---

[21] Since Strezhnev and Voeten's (2013) ideal points range from -3 to 3, the theoretical range of Divergence$_{L,P}$ is $[0, 6]$, and since the sample maximum is ≈ 5.1, there is no threat of truncation that would require a different estimator, such as tobit (Tobin 1958).

[22] For a more detailed discussion of the United States as a coalition-builder and partner, see Chapter 6.

TABLE 3.2. *Probit Model of Coalition Formation, 1946–2001*

$\text{Pr}(\text{Join}_{L,P} = 1)$	
Variable	**Estimates**
– Theoretical variables –	
$\text{Divergence}_{L,P}$	$-0.15\ (0.06)$***
CINC_P	$3.99\ (1.18)$***
$\text{Divergence}_{L,P} \times \text{CINC}_L$	$-0.23\ (0.54)$
– Control variables –	
CINC_L	$4.91\ (1.38)$***
CINC_T	$-1.83\ (1.74)$
$\text{Contiguity}_{P,T}$	$0.72\ (0.11)$***
$\text{Contiguity}_{L,T}$	$0.20\ (0.18)$
$\text{Allies}_{L,P}$	$0.51\ (0.11)$***
Democracy_L	$-0.05\ (0.20)$
Democracy_P	$0.18\ (0.10)$*
Democracy_T	$-0.01\ (0.18)$
United States_L	$-0.13\ (0.32)$
UNSC Support_L	$0.59\ (0.34)$*
Post-Cold War	$0.27\ (0.17)$
Intercept	$-3.10\ (0.22)$***
Model Statistics	
N	14158
Log-likelihood	-616.25
$\chi^2_{(\text{d.f.})}$	$147.90_{(14)}$***

Significance levels :   * : 10%   ** : 5%   *** : 1%

and P is positively associated with coalition formation, indicating that allies are an attractive pool from which to draw when choosing coalition partners. Notably, however, the significance of $\text{Allies}_{L,P}$'s coefficient does not interfere with that of the core theoretical variables, confirming empirically that alliances are neither necessary nor sufficient for the formation of coalitions.

Turning to the variables of primary interest, the theoretical model predicts that states will on average prefer partners that both offer valuable military assistance (Hypothesis 3.1) and share their preferences (Hypotheses 3.2), yet the discouraging effect of preference divergence should be smaller in magnitude for more powerful partners (Hypothesis 3.3). The coefficients on both $\text{Divergence}_{L,P}$ and CINC_P ostensibly bear this out; both are in the expected direction, negative and positive respectively, and both coefficients are distinguishable from zero at the $p < 0.01$ level. However, since these terms are also components of an interaction term, their coefficients can only be interpreted directly as the estimated effect when the other component term is zero. In other

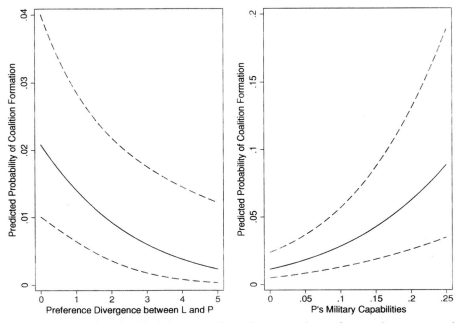

FIGURE 3.2. Predicted probability of coalition formation by preference divergence and potential partner's military capabilities with 95% confidence intervals

words, preference divergence discourages coalition formation when potential partners offer very little in the way of military benefits, while partner capabilities encourage coalition formation when leader and partner have identical preferences. While consistent with the theory's predictions, these coefficients are not the best indicator of the independent effects of these particular variables.

A more appropriate way to assess the effects of the component terms is to consider their effects on an "average" leader-partner dyad, which Figure 3.2 presents graphically using simulations based on the estimated coefficients.[23] In the left panel, lead states are most likely to take on partners with preferences similar to their own – that is, when Divergence$_{L,P}$ is low – and decreasingly willing to as preferences diverge. Preference diversity poses some unique challenges for choosing bargaining strategies, as discussed further in Chapter 4, but for now it is worth noting that Benson (2012, p. 138) shows that preference divergence in alliances is associated with ambiguous commitments, precisely because states are unwilling to pay the costs of committing to the defense of those partners most likely to drag them into undesirable wars. On the other hand, in the right panel, states are increasingly likely to build a coalition as

---

[23] All continuous variables are held at their medians, while all dummy variables are held at their modes.

the potential partner grows more powerful. The confidence in the relationship decreases as potential partners grow in strength, as *very* few states score above $CINC_P = 0.1$, which equates to controlling 10% of global military capabilities in a given year. Lead states thus appear less likely to form a coalition with a given partner as their preferences diverge but more likely to take on a particular state as its military capabilities grow ever larger.

Hypothesis 3.3 also predicts that the effect of preference divergence – that is, the size of its coefficient – should diminish as the military power of a potential partner increases, because lead states should be willing to make ever larger political concessions to secure the cooperation of such valuable partners. This implicates the interaction term (Divergence$_{L,P}$ × CINC$_P$). However, while the fact that it fails to achieve significance is consistent with the prediction that the effects of divergent preferences diminish as a potential partner grows more powerful, interpreting interaction terms in a limited dependent variable model is not straightforward, particularly when both interactive variables are continuous (Ai and Norton 2003, Brambor, Clark, and Golder 2006).

How one chooses to analyze interaction effects in the limited dependent variable context depends, in large part, on the question one is asking. While the interaction term itself reflects the estimated effect of a joint increase in the components of the outcome variable (Ai and Norton 2003) – that is, increasing divergence and increasing military power – the specific terms of Hypothesis 3.3 suggest a subtly different interpretation. Specifically, I expect that the marginal effect of preference divergence, moving from low to high levels of the variable, should be smaller at higher values of P's capabilities than at lower levels. To assess this hypothesis, I estimate the marginal effect of preference divergence across the range of possible values of military capabilities, then plot the estimated marginal effects against CINC$_P$, using the method proposed by Brambor, Clark, and Golder (2006) for limited dependent variables.

Figure 3.3 plots the marginal effect of leader-partner preference divergence across a wide range of partner military capabilities. Specifically, the vertical axis plots the change in the predicted probability of coalition formation as the result of moving from near-perfect preference similarly, or Divergence$_{L,P} = 0.05$ to near-perfect preference divergence, or Divergence$_{L,P} = 2.5$, as the potential partner's military capabilities range from virtually nil to nearly 40% of the total military capabilities in the system, a level achieved only by the United States in this sample.[24] As indicated by the negative marginal effect, increasing preference divergence decreases the probability of coalition formation when a potential partner brings little to the table militarily, but the marginal effect of preference diversity diminishes as the partner becomes more attractive militarily, as predicted by Hypothesis 3.3. In fact, as soon as a potential partner wields roughly 25% of global military capabilities – or as it enters the club of

---

[24] As in Figure 3.2, all variables are held at their means or modes as appropriate.

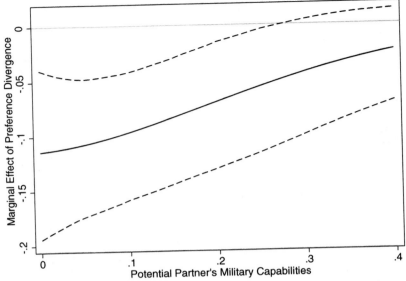

FIGURE 3.3. Marginal effect of dyadic preference divergence by potential partner's military capabilities with 95% confidence intervals

the largest military powers – then increasing preference diversity has *no* statistically discernible impact on the probability of coalition formation between a lead state and a potential partner. Thus, it appears that states prefer to minimize the concessions they must make to ensure cooperation, such that they favor forming coalitions with states that share their foreign policy preferences. However, they become less selective in their choice of partner when it offers a substantial boost to their military power.

Lastly, the coefficient on L's military capabilities, $CINC_L$, is both positive and statistically significant. This is inconsistent with the theoretical prediction of Proposition 3.3, which has stronger lead states less willing to build coalitions than weaker ones. However, the positive relationship between L's capabilities and crisis participation derived from Proposition 3.8 should induce caution in interpreting this coefficient's sign or its significance; it is not clear ex ante which effect – strong states' limited need for assistance or their overrepresentation in the sample of crisis participants – should dominate in a sample of observed crises. The Heckman probit with sample selection characterized in Model 2 of Table 3.4 in the appendix shows that strong states *are* more likely to enter crises in the first place, although the lack of a strong correlation in errors across selection (crisis) and outcome (coalition formation) stages prevents us from drawing effective inferences about the effect of L's capabilities on coalition formation.

TABLE 3.3. *OLS Model of Preference Diversity in Coalitions, 1946–2001*

Variable	Divergence$_{L,P}$ Estimates
– Theoretical variable –	
$CINC_P$	3.81 (1.47)**
– Control variables –	
$CINC_L$	−2.28 (0.87)**
$CINC_T$	−4.45 (2.19)**
Contiguity$_{P,T}$	0.03 (0.14)
Allies$_{L,P}$	−0.92 (0.22)***
United States$_L$	1.33 (0.32)
UNSC Support$_L$	0.35 (0.36)
Post-Cold War	−0.13 (0.38)
Intercept	1.32 (0.20)***
Model Statistics	
N	130
Model F	10.39 $_{(8,50)}$***

Significance levels : * : 10%   ** : 5%   *** : 1%

However, in the appendix I also estimate a simpler model of the probability that L forms a coalition with *any* partner (Model 1 in Table 3.4). This dyadic dataset measures only features of L and T, including a reduced-form measure of L's chances of defeating T militarily, or

$$p_L = \frac{m_L}{m_L + m_T} \equiv \frac{CINC_L}{CINC_L + CINC_T},$$

matching that found in the theoretical model. Notably, an increase in this variable *is* associated with a statistically significant ($p < 0.01$) decrease in the probability that L forms a coalition. Therefore, those states that most need coalition partners to boost their military prospects – that is, those with smaller shares of the total military capabilities in the principal dyad – are also those most likely to build coalitions.

3.3.4 Results: Coalition Diversity

The final empirical model, presented in Table 3.3, focuses on the 130 pairs of observed leader-partner dyads that go on to form coalitions in the data. The outcome variable is the observed difference in UNGA ideal points between lead states and their partners – specifically, Divergence$_{L,P}$ from the probit model estimated earlier (Strezhnev and Voeten 2013). Hypothesis 3.4 states that the divergence in preferences between those states that do form coalitions should

.......se in the potential partner's military capabilities, and this relationship is borne out by the coefficient on $CINC_P$, which is in the expected direction and statistically distinguishable from 0 at $p < 0.05$. Stronger partners are thus associated with more diverse coalitions due to L's willingness to make larger side payments for larger contributions.

Among the control variables, three achieve statistical significance. First, consistent with the expectations of the theoretical model, stronger lead states are associated with a significantly lower diversity of preferences in their coalitions, because they can afford to be choosier with those states they wish to compensate. Second, an increase in the target's military capabilities is associated with a *reduction* in the diversity of observed coalition preferences, which is unsurprising given the complicated conditional relationship between target strength and coalition formation found in the theoretical model. Third, the presence of an alliance tie between L and P is associated with a reduced level of preference divergence among coalition partners, indicating that coalitions built around alliance partnerships will be uniquely homogeneous in their preferences.

The natural interpretation of OLS coefficients is that they represent the expected change in the outcome variable for a unit change in the independent variable (Angrist and Pischke 2009, ch. 3), but since the variables in Table 3.3 are scaled differently, assessing their substantive impact is not straightforward. To aid in judging just how much the partner's military capabilities affect preference diversity in coalitions, I simulate two "average" leader-partner pairs in Figure 3.4 that are identical in all respects save one: whether the pair of states in question are allied.[25] In the left panel, potential partners range from nearly powerless to maximally powerful, and as partners grow stronger, they are on average more divergent from lead states in their military preferences. When the partner is virtually powerless, the predicted divergence in preferences is about 1.1, which is roughly equivalent to the divergence between Greece and Greek Cyprus in their crisis (ICB #223) against Turkey in 1967. On the other hand, when the partner reaches the level of a superpower at the right end of the scale, predicted divergence is about 2.5, which is similar to the divergence in preferences between the United States and Zaire when the former sided with the latter against Angola in 1975 (ICB #260).

The right panel presents the same relationship between partner capabilities and preference divergence between leader and partner, but in this case L and P share a preexisting treaty of alliance in the form of a defense pact or entente (Gibler and Sarkees 2004). As indicated by the coefficient on $Allies_{L,P}$ in Table 3.3, allied states on average build coalitions with more similar preferences than nonallied states, but the partner's capabilities still exercise a strong effect on the choice of coalition partner. However, the most diverse predicted allied coalition is roughly equivalent to the least diverse predicted nonallied

---

[25] As with previous simulations of predicted outcomes, all continuous variables are held at their means, dichotomous variables at their modes.

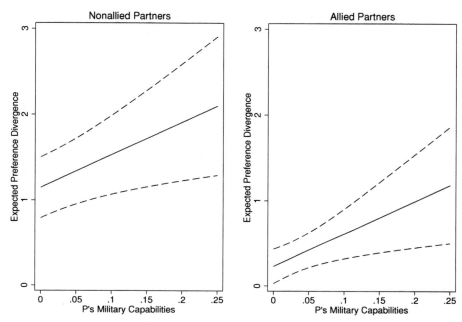

FIGURE 3.4. Expected preference divergence between leader and partner by allied status and potential partner's military capabilities with 95% confidence intervals

coalition – about as aligned as the United States and France were in the Berlin Crisis of 1958 (ICB #168). The least predicted divergence between allied leader and partner is about 0.19, which reflects the tightly aligned pairing of the United Kingdom and France in the Suez Crisis of 1956 (ICB #152). Thus, while a partner's military value can go a long way to explaining the presence of states with highly divergent preferences inside coalitions, it appears that coalitions made of allies see overall less divergence among their members – which, as discussed in Chapters 4 and 5, can have serious implications for both the probability of war and the expansion of conflicts.

3.4 APPLICATION: TURKEY AND THE IRAQ WARS

The preceding discussion can also shed some light on the puzzle posed at the beginning of the chapter: Why was Turkey a part of the 1991 Gulf War coalition yet mostly sidelined during the 2003 Iraq War, despite consistent regional interests and a similar potential military contribution – the ability to threaten Iraq with a two-front war – in each case? The answer turns on the changing value of Turkey's military contribution relative to the concessions required to secure it; while the United States found Turkey easily compensated in 1991, leading to a simple exchange of cooperation for a package of political concessions, a host of factors would render its cooperation more expensive in 2003, leading ultimately to a failed attempt to negotiate the terms of its coalition membership.[26]

After Iraq's attempted annexation of Kuwait in August 1990, the United States set about building a military coalition aimed first at deterring a further move into the northeastern Saudi oil fields and, more importantly for this story, coercing Iraq to withdraw from Kuwait under pain of a threatened war.[27] Armed with the first explicit UN Security Council approval to use military force since North Korea's invasion of its southern neighbor in 1950 (see Voeten 2005), the coalition featured promised military contributions from a wide range of interested states, including Saudi Arabia, the United Kingdom, France, Syria, Egypt, and, of course, Turkey.[28] Despite the size of the coalition arrayed against it, Iraq refused to evacuate Kuwaiti territory, and Operation Desert Storm began on 17 January 1991 with a lengthy air campaign designed to prepare the way for the invasion of Kuwait and southern Iraq (Atkinson 1993).

Once war broke out, Turkey's role in the coalition was twofold, allowing the use of its main NATO airbase, Incirlik, to manage the northern part of the air campaign and placing "nearly 100,000 Turkish troops near the Iraqi border, pinning down an Iraqi force of equal size" with the mere threat of invasion from the north (Sayari 1992, p. 14). Forcing such a division of the Iraqi army, then touted as the fourth largest in the world and battle-hardened after eight years of war with Iran (Atkinson 1993, p. 4), facilitated coalition efforts in the south, where Iraqi forces were dislodged from Kuwait with surprising ease; the ground campaign lasted roughly 100 hours before the signing of a ceasefire that restored Kuwaiti sovereignty and brought to a halt the destruction of Iraqi military hardware during the retreat from Kuwait City along the so-called "highway of death" (Freedman and Karsh 1991, p. 34). The terms of the ceasefire satisfied the coalition's primary war aims of the restoration of the Kuwaiti royal family and the degradation of Iraq's conventional military capabilities, reducing in principle its ability to attack or threaten its southern neighbors in the future.

---

[26] Note that failed coalition negotiations occur only "off the equilibrium path" in the theoretical model, due to the complete-information character of the game. However, the model still tells us *why* negotiations would fail if failure were to occur – say, because of some form of uncertainty – ensuring that it can be used to shed light on the puzzle of Turkey in 1991 and 2003.

[27] Recall that, in the data presented in Chapter 2, Kuwait is considered the lead state at the outset of the crisis, when it requested American assistance in pressuring Iraq to leave Kuwait. However, as Operation Desert Shield turned into Operation Desert Storm, the United States took a lead role in building a coalition to eject Iraqi forces from Kuwait, at which point it can be considered the lead state in this application of the theory.

[28] Freedman and Karsh (1991) correctly note that the former countries put boots on the ground in Iraqi or Kuwaiti territory, but my definition of military cooperation, given in Chapter 2, is broader, encompassing the basing and staging rights granted by both Saudi Arabia and Turkey.

Getting Turkey on board, however, was not a trivial matter. Though President Turgut Özal was decidedly more pro-American and less inclined toward regional neutrality than his predecessors were (Sayari 1992, p. 18), convincing a military establishment and a public historically "reluctant to allow the United States use of its territory as a staging ground in non-NATO contingencies" (Kuniholm 1991, p. 39) would prove challenging.[29] Iraq was, after all, a major trading partner likely to remain so after any war (*ibid.*, p. 40), and destabilizing the northern parts of the country – home to a large, dissatisfied Kurdish population – raised concerns about Turkey's own restive Kurdish minority. Alliance ties to the United States notwithstanding, then, Turkey would have to be compensated handsomely for joining the coalition:

> In return . . . Özal anticipated that the United States would more readily fulfill some of Turkey's expectations: the expansion of trade relations between the two countries, especially greater access to US markets for Turkish exports; increased US military assistance for the modernization of the Turkish armed forces; and the development of a "new strategic relationship" between Turkey and the United States. (Sayari 1992, p. 14)

Özal also hoped for American diplomatic assistance in Turkey's attempts to join the European Community, as well as an adjustment in the ratio of military aid disbursed between Turkey and its chief rival, fellow NATO member Greece (Kuniholm 1991, p. 35).

Although Turkey would go on to express dissatisfaction with the financial terms of the bargain years later (Önis and Yilmaz 2005, Robins 2003), it received many of its desired concessions in return for the use of its airbases and the deployment of its troops to the Iraqi border. After the war, the United States "doubled the value of Turkey's textile quota," "provided . . . $282 million in additional military and economic assistance in 1991," persuaded Egypt to buy a Turkish-made version of the F-16 fighter jet, and hammered out agreements with the Gulf states to help Turkey recover some of the billions of dollars lost in the crisis due to military expenditures and disrupted regional trade (Sayari 1992, p. 19). In sum, the United States decided that Turkey's military contribution to the 1991 Persian Gulf War was worth the price (Bush and Scowcroft 1998), then compensated it for its participation in a massive coercive effort that might have proven both domestically unpopular for the Turkish government and quite costly – politically, financially, and militarily – if the war were to drag on through the summer or draw in other states (Freedman and Karsh 1991, p. 8).

Twelve years later, American forces led another coalitional invasion of Iraq, this time with the explicit aim of toppling the Ba'athist government of Saddam Hussein. Concerned that Iraq could not credibly commit not to develop nuclear, chemical, and biological weapons – and deeply skeptical of claims that Iraq had

---

[29] Despite the involvement of NATO allies, this was not a NATO operation, since it neither invoked the collective defense clause nor saw the involvement of the alliance's institutions.

no such active programs – the United States decided to solve the problem with forcible regime change (Debs and Monteiro 2014, Reiter 2009). However, while the 1991 coalition, buoyed by explicit UNSC support, was unusually large – tied for the largest in the coalitions data, in fact – the 2003 coalition was considerably narrower: the United States and the United Kingdom were the only states contributing significant military forces, while Kuwait provided the staging ground for the invasion.[30] American decision makers, however, had wished to attack on two fronts, with one force moving up from Kuwait and another pressing down from Turkey in the north. A second front would have forced Iraq to divide its army once again, facilitating its collapse by reducing its ability to concentrate defensive firepower on a single front (Keegan 2005, ch. 6).[31]

With Saudi support limited to allowing command and control – but not combat – operations at its Prince Sultan Air Base (CNN 2003), Turkey's potential military contribution was arguably more valuable in 2003 than it was in 1991, which should, all else being equal, have rendered the United States more willing to compensate it. However, the two countries failed to reach an agreement over the terms of that cooperation, as Turkey's parliament rejected by a narrow margin a proposal to allow thousands of American troops to enter the country and open the northern front (Önis and Yilmaz 2005, p. 276). Why, given Turkey's potentially valuable military contribution – as well as its own desire to establish order in northern Iraq and pursue Kurdish insurgents (Gordon and Trainor 2006, p. 42) – did it fail to reach an agreement with the United States in 2003?

To answer this question, the model points us toward factors that would cause Turkey's required compensation to outstrip the American willingness to pay for it. Indeed, several critical factors had changed in Turkey by 2003. To be sure, the war was deeply unpopular with the public (Robins 2003, p. 560), with the lack of UNSC approval for the much-discussed "second resolution" authorizing force serving as a lighting rod for criticism. Turkey was also a more democratic country, sensitive to public opinion in a way it was not in 1991. Further, both civilian and military officials were concerned about the "emergence of a Kurdish state in northern Iraq"; the future status of Iraqi Turkomans, historically supported as balancers against the Kurds in Kirkuk;

---

[30] The so-called "coalition of the willing," largely made up of states that had voiced public support for the American operation, was of course much larger – at one point reaching a claimed forty-six states (Leetaru and Althaus 2009) – but given my focus on military contributions, cheap-talk expressions of support fall outside the scope of the analysis. Then-President Bush's protestations notwithstanding, I can safely defend myself against the charge that I "forgot Poland."

[31] As it happened, the Iraqi military collapsed, oftentimes disappearing into the population, with wholly unexpected swiftness, defying prewar predictions that, at a minimum, it would put up stiff resistance around major cities, especially Baghdad (Keegan 2005, ch. 1).

"the potential for economic losses; dread of a humanitarian crisis in Iraq and along the Turkey-Iraq border; and concern for effects on regional stability" (Migdalovitz 2003, p. 7). Where the relatively limited aims of the 1991 war did not portend such large strategic costs for Turkey, the wholesale replacement of the Iraqi government risked a host of more serious problems, in terms of both geopolitics and domestic electoral politics (Önis and Yilmaz 2005, p. 275). In other words, Turkey's cooperation in 2003 would come with a larger price tag than it did in 1991, including reassurances that Iraqi territorial integrity would be preserved, the freedom for Turkish forces to establish order 40 kilometers into Kurdish northern Iraq, and $90 billion in aid (Hale 2007, pp. 125, 130).

If Turkey's demands were larger, we must also answer the question of why the United States was unwilling to meet them, despite what ostensibly should have been a greater willingness to adjust for the lack of Saudi support so critical to the invasion of 1991. The United States wanted to deploy 62,000 troops to Turkey, tasked with invading from the north and achieving the division of Iraqi forces and attention, and it acceded easily to one of Turkey's primary requirements: that of not dismembering Iraq (Migdalovitz 2003, p. 12). Further,

the US was prepared to put together a financial package by way of compensation for the cost of the conflict; allow Turkish troops to take part in any military operations in northern Iraq; set up a missile defence umbrella to protect strategic sites in Turkey, from dams to oil pipelines; and provide a package of military benefits, including upgrading Turkey's bases and supplying specialist reconnaissance and intelligence technology. (Robins 2003, p. 561)

Further, as the specifics of these terms were discussed, the United States also made a diplomatic push on Turkey's potential accession to the European Union at the Copenhagen Summit in November 2002, which Turkey had for some time considered necessary before any commitment to the coalition (Önis and Yilmaz 2005, pp. 273, 276).

However, the terms of this package in several cases fell short of what the Turkish government felt it needed before consenting to participate in a war so unpopular with the public; the United States would approve only a limited Turkish "security belt" in northern Iraq (Hale 2007, p. 125) – insisting that Turkish troops act in coordination with its own (Robins 2003, p. 564) and in a noncombat role (Migdalovitz 2003) – and offered only $6 billion in grants and $10 billion in loan guarantees (Hale 2007, p. 133).[32] Ultimately, this proposed deal failed to overcome widespread public and legislative opposition within

---

[32] See also Migdalovitz (2003), who gives the same numbers for financial assistance but also indicates that the offer included enhanced trade as well, as well as Keegan (2005, p. 138) for a characteristically sardonic take on the failed negotiations.

Turkey, failing in parliament on 1 March 2003, mere weeks before the start of the war, even as tens of thousands of American troops sat waiting on ships in the Mediterranean off the Turkish coast (Migdalovitz 2003).

Both governments had been optimistic, if guardedly so, about the proposal's parliamentary chances, and its failure was a mild surprise (Önis and Yilmaz 2005, p. 276), which Foreign Minister Yasir Yaks attributed to Turkey's belief that it could hold out for a better deal, given its potentially significant military contribution (Migdalovitz 2003, pp. 12, 13). Turkey would eventually grant overflight rights for coalition aircraft, but after the proposal's initial failure in the legislature, the United States made no effort to adjust the terms of the deal, even as Turkish officials expressed a desire to revisit elements of a memorandum of understanding about a Turkish presence in northern Iraq (Önis and Yilmaz 2005, p. 276). Why were no adjustments made to the package in hopes of securing cooperation? Ultimately, American decision makers were divided on the military necessity of a second front, especially as the Iraqi military had weakened significantly since the end of the last war. Secretary of State Colin Powell, for instance, thought that forcing Iraqi units into a single front would make them more vulnerable to overwhelming American firepower, hastening their collapse in a way that even dividing their forces might fail to achieve (Gordon and Trainor 2006, p. 115).

In the final accounting, the United States decided that Turkey's military contribution was not worth the terms required to ensure its cooperation, particularly the prospect of Turkish forces establishing order – that is, pursuing Kurdish insurgents and perhaps creating problems for the occupation – deep into northern Iraq (Gordon and Trainor 2006, Hale 2007). In terms of the model, Turkey's required compensation increased from 1991 to 2003, thanks to what it believed to be a valuable military contribution. However, given the degradation of the Iraqi military since 1991 and fresh optimism over the small-footprint strategy practiced in the invasion of Afghanistan in 2001 (Gordon and Trainor 2006, Keegan 2005), the American willingness to meet Turkey's terms had not increased as quickly as the terms themselves. There was also a smaller public component to the stakes; where the 1991 war implicated global territorial norms and the preservation of the international order, the 2003 war was concerned more narrowly with who would govern Iraq, and any demands Turkey made over postwar spoils and influence would surely be relatively more expensive for the United States to satisfy. This relatively larger private component, combined with American assessments of a small military contribution, makes the United States' unwillingness to pay a higher price understandable, consistent with the logic of Proposition 3.6. If the United States could secure cooperation on the cheap, limiting Turkey's military role and making a smaller financial side payment while still promising not to dismember Iraq, then such cooperation would have been welcome. As it was, the price necessary to secure cooperation simply proved too high, and the invasion of Iraq would be launched along a single, southern front from Kuwait, reaching Baghdad, toppling the government, and forcing Iraqi President Saddam Hussein into hiding in a matter of weeks.

3.5 SUMMARY AND DISCUSSION

George Kennan noted after the Second World War that, for the Allies,

> There was no prospect of victory over Germany, unless it were with the help of Russia. But for such help, even if it were forthcoming, the Western democracies would have to pay heavily in the military consequences of the war and in the demands that would be raised at the peace table. Their military purposes, in other words, were mortgaged in advance. (Kennan 1984, p. 77)

The theoretical and empirical models in this chapter suggest that Kennan's intuition about the workings of the Allied coalition was correct and of broader applicability: a complex interplay of military power and foreign policy preferences are important factors in the process of coalition formation.

Specifically, when building coalitions, states seek to increase the military power they can bring to bear against their enemies while minimizing the concessions necessary to ensure their partners' cooperation. This produces a fundamental trade-off between increasing one's chances of military success and the costs of compensating partners for their assistance. When potential partners offer only modest increases in the chances of success, lead states may forego their cooperation if their preferences are sufficiently divergent, but when the military benefits of cooperation are substantial, states become less selective and will take on partners with preferences increasingly divergent from their own. This basic, transactional story also offers a plausible account of why the United States and Turkey were able to reach an accommodation over military cooperation for the Gulf War of 1991 but were unable to do so for the Iraq War of 2003; while the price of Turkish cooperation had increased from war to war, American willingness to pay the required price had not kept pace.

Finally, empirical models of coalition formation in late twentieth and early twenty-first-century crises, as well as preference divergence inside those coalitions, show that coalition-builders tend to pursue partners with preferences similar to their own, but only when such partners offer limited military contributions. When partners offer more significant military assistance, preference divergence makes less of an impact on lead states' calculations, and it leads to coalitions with a greater divergence of preferences. Coalitions are, indeed, built first and foremost to "win" (cf. Riker 1962), but the price of winning is often prohibitive. As discussed in Chapter 2, the choices that states make in light of the trade-off between aggregating power and minimizing concessions to their partners have implications for both (a) whether crises escalate to war or end peacefully and (b) whether conflicts expand by provoking balancing behavior by outsiders to the original conflict. These processes are the subjects of Chapters 4 and 5, respectively.

3.6 APPENDIX

3.6.1 Proofs

Proof of Proposition 3.1. Begin with P's response to L's proposal s, which P accepts if

$$EU_P(\Gamma, LP) > u_P(\text{reject}) \Leftrightarrow p_{LP}(\beta\gamma + (1-\gamma)(1-s)) - c_P \geq p_L\beta\gamma,$$

or when $s \leq \hat{s}$, where

$$\hat{s} \equiv \frac{c_P + \beta\gamma p_L + (-\beta\gamma + \gamma - 1)p_{LP}}{(\gamma - 1)p_{LP}}.$$

Otherwise, it rejects. Now consider L's coalition proposal if it selected into the crisis. If L wishes to induce P's acceptance, it proposes $s^* = \hat{s}$, which ensures that it makes no unnecessary compromises. L proposes such a coalition when

$$EU_L(\Gamma, LP) > EU_L(\Gamma, L) \Leftrightarrow p_{LP}(\gamma + s^*(1-\gamma)) - c_L > p_L - c_L,$$

or when

$$\beta > \frac{c_P - (p_{LP} - p_L)}{\gamma(p_{LP} - p_L)} \equiv \beta^\dagger.$$

Otherwise, it acts unilaterally. Finally, consider L's choice over initiating a crisis. When $\beta \leq \beta^\dagger$ such that L acts unilaterally, it initiates a crisis when $EU_L(\Gamma, L) > u_L(q) \Leftrightarrow p_L - c_L > q$, or when $q < p_L - c_L \equiv q_L$; otherwise, it tolerates the status quo. When $\beta > \beta^\dagger$ such that L forms a coalition, it initiates a crisis when $EU_L(\Gamma, LP) > u_L(q) \Leftrightarrow p_{LP}s^* - c_{LP} > q$, or when

$$q < p_{LP} - c_L - c_P + \beta\gamma(p_{LP} - p_L) \equiv q_{LP}.$$

Otherwise, it tolerates the status quo.   □

Proof of Proposition 3.2. The claim follows directly from Proposition 3.1.   □

Proof of Proposition 3.3. To verify that $\beta^\dagger$ increases in m_L, note that the first derivative with respect to m_L is

$$\frac{\partial\beta^\dagger}{\partial m_L} = \frac{c_P(2m_L + m_P + 2m_T)}{\gamma m_P m_T} > 0,$$

which is strictly positive.   □

Proof of Proposition 3.4. To verify that $\beta^\dagger$ decreases in m_P, note that the first derivative with respect to m_P is

$$\frac{\partial\beta^\dagger}{\partial m_P} = -\frac{c_P(m_L + m_T)^2}{\gamma m_P^2 m_T} < 0,$$

which is strictly negative.   □

Proof of Proposition 3.5. To verify the claim about the relationship between $\beta^\dagger$ and m_T, note that the first derivative with respect to γ is

$$\frac{\partial \beta^\dagger}{\partial m_T} = \frac{c_P \left(m_T^2 - m_L \left(m_L + m_P \right) \right)}{\gamma m_P m_T^2}.$$

This quantity is negative when $m_T < \sqrt{m_L \left(m_L + m_P \right)}$, and it is weakly positive when $m_T \geq \sqrt{m_L \left(m_L + m_P \right)}$. $\qquad\square$

Proof of Proposition 3.6. To verify the claim about the relationship between $\beta^\dagger$ and γ, note that the first derivative with respect to γ is

$$\frac{\partial \beta^\dagger}{\partial \gamma} = \frac{1}{\gamma^2} \left(1 - \frac{c_P}{p_{LP} - p_L} \right).$$

This quantity is negative when $p_{LP} - p_L < c_P$, and it is weakly positive when $p_{LP} - c_L \geq c_P$. $\qquad\square$

Proof of Proposition 3.7. To verify that q_{LP} increases in m_P, note that the first derivative with respect to m_P is

$$\frac{\partial q_{LP}}{\partial m_P} = \frac{m_T \left(1 + \beta \gamma \right)}{\left(m_L + m_P + m_T \right)^2} > 0,$$

which is strictly positive. However, since

$$\frac{\partial^2 q_{LP}}{\partial m_P \partial m_L} = - \frac{2 m_T \left(1 + \beta \gamma \right)}{\left(m_L + m_P + m_T \right)^3} < 0,$$

the effect diminishes as m_L increases. $\qquad\square$

Proof of Proposition 3.8. To verify that q_L increases in m_L, note that the first derivative with respect to m_L is

$$\frac{\partial q_L}{\partial m_L} = \frac{m_T}{\left(m_L + m_T \right)^2} > 0,$$

which is strictly positive. Next, to verify the claim about the relationship between q_{LP} and m_L, note that the first derivative with respect to m_L is

$$\frac{\partial q_{LP}}{\partial m_L} = \frac{m_T \left(2 m_L \left(m_T - \beta \gamma m_P \right) + m_L^2 - \beta \gamma m_P^2 - 2 \beta \gamma m_P m_T + m_T^2 \right)}{\left(m_L + m_T \right)^2 \left(m_L + m_P + m_T \right)^2}.$$

This quantity is positive when

$$\beta < \frac{\left(m_L + m_T \right)^2}{\gamma m_P \left(2 m_L + m_P + 2 m_T \right)} \equiv \hat{\beta},$$

and it is weakly negative when $\beta \geq \hat{\beta}$. $\qquad\square$

Proof of Proposition 3.9. To verify that q_L increases in m_T, note that the first derivative with respect to m_T is

$$\frac{\partial q_L}{\partial m_T} = -\frac{m_L}{(m_L + m_T)^2} < 0,$$

which is strictly negative. Next, to verify the claim about the relationship between q_{LP} and m_T, note that the first derivative with respect to m_T is

$$\frac{\partial q_{LP}}{\partial m_T} = \frac{\beta\gamma m_L}{(m_L + m_T)^2} - \frac{(\beta\gamma + 1)(m_L + m_P)}{(m_L + m_P + m_T)^2}.$$

This quantity is positive when

$$\beta > \frac{(m_L + m_P)(m_L + m_T)^2}{\gamma m_P \left(m_L(m_L + m_P) - m_T^2\right)} \equiv \tilde{\beta}$$

and when either (a) $m_T \leq m_L$ or (b) $m_L + m_P > m_T^2/m_L$ and $m_T > m_L$. Otherwise, if any set of conditions is not met, it is weakly negative. □

Proof of Proposition 3.10. Solving $q_{LP} > q_L$ for β yields

$$\beta > \frac{c_P - (p_{LP} - p_L)}{\gamma(p_{LP} - p_L)} \equiv \beta^{\dagger},$$

which is the condition for coalition formation identified in Proposition 3.1. □

3.6.2 Assessing Potential Selection Bias

As discussed earlier, the expectation of coalitional support can encourage states to initiate crises they might otherwise avoid. While this implication is interesting in its own right, it also suggests that sampling on observed crises might be problematic if unobserved factors drive both processes, leading to correlations in the errors across (a) selection into crises and (b) the formation of coalitions within them (see Heckman 1979). If leaders are incentivized to initiate crises when they expect coalitional support, then variables that predict formation in this chapter's empirical models might lead to faulty inferences if the errors across both stages of the process are correlated. It is of course plausible that multiple processes of selection might be operative; states might avoid initiating crises against those states from which they expect to gain coalitional support, which would have the opposite, potentially offsetting effect on the observed sample. Nonetheless, to check the soundness of the decision to sample on crises, I estimate an empirical model below that checks for any correlations in the errors across crisis onset and coalition formation that might threaten inferences drawn in the chapter's main analyses.

Specifically, I estimate a probit model with sample selection, or Heckman probit, where the first stage predicts the onset of an ICB crisis (Wilkenfeld and Brecher 2010) between all directed pairs (dyads) of states in the international system (COW 2011) from 1946 to 2001, then estimates the extent to which errors are correlated across the equations with the parameter ρ. In other words, the selection variable Crisis$_{L,T}$ = 1 if the ordered pair of states enters the sample of crises in a given year and zero otherwise. The second, or outcome, stage indicates whether state L takes on a coalition partner during the crisis, such that Coalition$_L$ = 1 if L forms a coalition with at least one partner and zero otherwise. Note that this implies a different unit of observation and outcome variable than the model of coalition formation in the main part of the chapter, which focused on leader-potential partner dyads. I do this in order to ensure comparability in units of observation across the selection and outcome stages – directed dyads to directed crisis dyads – and because more complicated selection models produced substantively similar results: the errors across the stages of crisis selection and coalition formation are not strongly correlated, nor is it possible to reject the null hypothesis that the errors across the equations are independent (i.e., that $\rho = 0$).

To see how I reach these conclusions, begin with the outcome stage modeling the formation of a coalition inside a crisis: that is, when state L forms a coalition with at least one state. I model this as a function of both L's and T's military capabilities (Singer, Bremer, and Stuckey 1972, Singer 1987), L's share of total capabilities in the dyad, whether the dyad is land-contiguous (Stinnett et al. 2002), and both Polity-IV democracy scores (Marshall and Jaggers 2009). Next, to model selection into crises in the first place, I include the variables above, with one additional variable, an instrument or exclusion restriction (Angrist and Pischke 2009, ch. 4), that is correlated with crisis onset but not with coalition formation. Several candidates revealed themselves, but the most promising instrument turned out to be an indicator of joint democracy in the dyad, which is strongly correlated with the onset of a crisis between L and T but not significantly related to whether L forms a coalition.

Table 3.4 presents the results of the selection model, with the selection stage at bottom and the outcome stage above it. Several results are immediately of note, in that powerful states, democracies, and pairs of neighbors are uniquely likely to experience crises, while pairs of democracies are uniquely unlikely to enter crises with one another. These results are broadly consistent with extant work on the onset of crises and disputes short of war (see Bennett and Stam 2004). In the outcome stage, only relative capabilities have a statistically significant effect on the probability that L builds a coalition, indicating that states find partners more attractive to take on when they begin the crisis at a military disadvantage.

Outcome equation aside, the most important pair of statistics are the estimated correlation in errors and the likelihood-ratio test for independent equations. First, the estimated correlation in errors across the stages is $\rho = -0.120$, and with a standard error of 1.013, this very weak correlation coefficient is statistically indistinguishable from zero at conventional levels; the 95% confidence interval is $(-0.973, 0.956)$. Second, a likelihood-ratio test of the null

TABLE 3.4. *Probit and Heckman Probit Selection Models of Crisis Onset and Coalition Formation, 1946–2001*

	Pr(Coalition$_{L,P}$ = 1|Crisis$_{L,T}$ = 1)	
Variable	**Model 1** *Probit*	**Model 2** *Heckman Probit*
	Outcome Equation	
CINC$_L$	7.04 (2.61)***	6.63 (4.69)
CINC$_T$	−4.40 (3.27)	−4.73 (4.11)
Relative Capabilities	−0.90 (0.33)***	−0.90 (0.34)***
Contiguous$_{L,T}$	0.09 (0.23)	−0.07 (1.36)
Democracy$_L$	−0.01 (0.01)	−0.01 (0.01)
Democracy$_T$	−0.02 (0.013)	−0.02 (0.01)
Intercept	−0.89 (0.26)***	−0.42 (4.07)
	Selection Equation	
CINC$_L$		3.20 (0.45)***
CINC$_T$		3.20 (0.45)***
Relative Capabilities		0.00 (0.06)
Contiguous$_{L,T}$		1.44 (0.04)***
Democracy$_L$		0.01 (0.00)***
Democracy$_T$		0.01 (0.00)***
Joint Democracy		−0.48 (0.08)***
Intercept		−3.63 (0.04)***
	Model Statistics	
N	398	771120
ρ		−0.120 (1.013)
LR test $\rho = 0$		$\chi^2_{(1)} = 0.01$, $p = 0.907$

Significance levels :   * : 10%   ** : 5%   *** : 1%

hypothesis of no correlation, $\rho = 0$, fails to reject the null, and quite strongly; the p-statistic for the χ_2 test suggests that the probability with which this correlation could be produced by chance is as high as 0.907.

The results of the Heckman model indicate, therefore, that there are no significant unmeasured factors contributing substantially to sample selection bias when (a) the analyst samples on observed crises and (b) accounts for factors such as disputants' military capabilities and domestic political environment. This increases confidence in the inferences drawn from Tables 3.2 and 3.3 in the main body of the chapter.

4

Cooperation, Signaling, and War

He's playing a game . . . He thinks we're unwilling to act. If he thought there was real steel in the threat he wouldn't get away with it.

Prime Minister Tony Blair
quoted in Richardson (2000, p. 146)

Even with its considerable military might and unique power projection capabilities, the United States almost always finds itself securing the assistance of coalition partners in international crises. However, this added military leverage often comes with disagreements over how to engage the target state – disagreements whose implications for successful coercion are unclear. As discussed in Chapter 2, the Berlin Crisis of 1961 saw the United States search for a signaling strategy that would reassure nervous allies that it "would not be rash," even though "the Soviet Union had to be persuaded that it just might be" (Freedman 2000, p. 93) in defense of Western rights in the divided city. In other words, the United States would have liked one audience to believe that it was willing to risk a costly war, even as it reassured another audience that any possible war would not be too costly. Similar problems plagued NATO efforts with respect to Kosovo in 1998–1999, as member states engaged in a lengthy debate over whether to threaten Serbia with a full-scale ground invasion or a relatively low-cost air campaign (Clark 2001, chs. 6, 7). In each case, concerns about undermining military cooperation produced ostensibly weak threats – a slow, limited mobilization in the former and a limited air campaign in the latter – that would seem to undermine statements of resolve. However, while the crisis in Kosovo escalated to war after an apparently weak threat, the Berlin Crisis did not, suggesting that the link between coalitions and the escalation of crises to war may not be as straightforward as the bivariate relationships uncovered in Chapter 2 or as suggested by the extant literature (Christensen 2011, Lake 2010/2011).[1]

This chapter seeks to resolve the puzzle by asking how coalition partners affect the credibility of threats and the probability of war. More practically, when are partners a hindrance to, and when might they facilitate, the peaceful settlement of disputes? To answer these questions, I analyze a model of how a lead state chooses military threats in the shadow of both a target's uncertainty *and* a potential partner's choice over providing military support in the event of war. In more standard two-player models, resolute types face a problem of credibly separating themselves from irresolute types that would like to bluff (see Fearon 1997, Slantchev 2005). However, the addition of a potential coalition partner can change these dynamics substantially. For example, the partner may differ from the leader in its preferred level of military mobilization, because domestic-political, geographical, or resource constraints induce a particular sensitivity to the costs of war. A coalition partner's government may be able to sell an air campaign, but not a ground war, to its constituents, such that it can commit to cooperating only in the event of smaller mobilizations – which also happen to be less credible signals of resolve. Thus, the uneven distribution of the costs of war within coalitions plays a central role in shaping their collective strategy going in to the crisis.

While a lead state would like to convince the target of its resolve with a large mobilization, it must be mindful of losing a potential partner's military support should it mobilize too much. In other words, solving the target's information problem by mobilizing for war may create a commitment problem for an erstwhile partner, putting military cooperation in jeopardy. While diverse preferences and the challenges of securing military cooperation play prominent roles in theories of extended deterrence and alliance formation (e.g., Benson 2012, Morrow 1994, 2000, Smith 1995), they remain absent from most models of signaling, and this chapter's model bridges these two literatures. In this way, the model builds on that of the previous chapter by (a) modeling explicitly divergent preferences over bargaining strategies within the coalition, giving them substantive weight in the context of crisis bargaining; and (b) adding to a model of intra-coalitional bargaining a crisis bargaining game with incomplete information and the possibility of sending credible – that is, believable – signals of resolve.

The theory shows that the trade-off between signaling resolve and preserving the cooperation of a coalition partner has divergent effects on the probability of war – effects that depend on both the distribution of preferences within the coalition and military power, the same factors that drove coalition formation in Chapter 3. First, when the target state is relatively strong, the promise of

---

[1] Portions of this chapter also appear in: Scott Wolford. 2014. "Showing Restraint, Signaling Resolve: Coalitions, Cooperation, and Crisis Bargaining." *American Journal of Political Science* 58.1:146–157.

retaining the partner's support reduces the probability of war by discouraging an irresolute lead state from bluffing, which would fail to solve the informational problem and tempt the target to risk war. Thus, the weak or moderated threats around which coalitions often coalesce (see Byman and Waxman 2002, Papayoanou 1997) can be a force for peace against strong targets. This also implies that acting unilaterally can serve as a signal of resolve under some conditions. Second, when the target is relatively weaker, a resolute lead state masks its type by choosing a *low* level of mobilization, retaining the partner's support at the cost of failing to signal its resolve, leaving the target's information problem unsolved and generating a risk of war that would not exist without the prospect of securing the partner's cooperation. Thus, securing a partner's military cooperation can lead to war where the absence of such cooperation would not. Further, coalition leaders have greater incentives to mask their resolve behind ineffective signals when they wish to secure the cooperation of powerful partners.

In the remainder of this chapter, I present and analyze the theoretical model, taking care to draw out the empirical implications over the probability of war given the inherent problem of observing states' resolve in situations – that is, crisis bargaining – where there are clear incentives for misrepresentation. I then analyze an empirical model that links the preference diversity inherent in coalitions to the escalation of crises to war through the distribution of military power and to differences in escalation rates across coalitions and singletons. Before concluding, I also discuss in detail the processes by which skittish coalition partners shaped American military threats during the Berlin Crisis of 1961–1962 and the Kosovo Crisis of 1998–1999, showing how the desire to ensure military cooperation discouraged a risky strategy of bluffing in the former, preserving peace, yet failed to demonstrate true resolve in the latter, resulting in the nearly eighty-day NATO air campaign of the Kosovo War.

4.1 THE CHALLENGE OF CREDIBILITY

The problem of credible communication during crises emerges from states' incentives to lie about their resolve, or willingness to fight, to induce their opponents to offer generous terms (cf. Fearon 1995). After all, if a claim that one is willing to fight will be believed, why not make just that claim and reap the rewards of changing an opponent's beliefs? This basic credibility problem often undermines the value of simple, cheap-talk diplomacy, because truly resolute states have difficulty revealing themselves (Jervis 1970), absent some cost attached to making threats of war. As a result, they may be unable to disabuse their opponents of the optimism required to propose miserly bargains that resolute states end up rejecting in favor of war – even though both sides would have preferred to strike a bargain to save the costs of fighting (Fearon 1995). Credibly demonstrating resolve can thus help states win larger concessions and avert war, but doing so requires combining a threat with some cost that would make bluffing painful, separating the truly resolute from the irresolute and thereby helping opponents fashion bargains that are more likely to be accepted (Fearon 1997, Slantchev 2005).

To think about the problem more formally, suppose that state L can simply say to state T through diplomatic channels that it is willing to fight in response to all but the most generous proposals. Then, suppose that T will believe this to be true. If this is the case, then T will try to avert a costly war by offering a generous bargain; however, if T will believe such a statement, and the statement carries no inherent cost, then L will claim to be resolute even if it is not, thereby undermining T's incentives to believe L's diplomatic statements in the first place. As a result, simple claims of resolve are typically credible only when they carry some cost, or some disincentive to lie about the willingness to fight (cf. Spence 1973); the costlier the signal, the more credible the statement, or the more it alters the target state's beliefs. In the bilateral context, national leaders can sink costs by increasing defense spending (Fearon 1997) or fighting today to cultivate reputation (Schelling 1966, Wolford 2007), tie their hands by generating domestic audience costs in the event of backing down by making public commitments (Fearon 1994, Tarar and Leventoğlu 2013), or mobilize the military, which performs both functions (Slantchev 2005, Tarar 2013).[2] However, states in multilateral crises are often subject to competing pressures from coalition partners to limit the costs of a potential war, even when increasing those potential costs might make threats more credible.

Coalitions, in the typical story, would seem to pose an acute challenge here; divergent preferences within coalitions are often blamed for such weak, diluted – or, more properly, insufficiently costly – signals (e.g., Bellamy 2000, Byman and Waxman 2002, Christensen 2011). Compared to a unilateral military mobilization, which increases the chances of victory but requires no concessions to the partners providing added capabilities, coalitions appear ready-made to produce dangerously watered-down threats and significant cooperative problems. However, the strategic linkages between multilateral threats and military cooperation are not so straightforward. The promises of third-party assistance in formal alliances can lend credibility to deterrent threats (Leeds 2003b, Morrow 1994, 2000), even as the potential reactions of third parties, partners in particular, have also been invoked to explain failures to signal resolve when doing so might have avoided war (see Benson 2012, Fearon 1997, Russett 1963). Games between coalition partners and against their adversaries are strategically linked (Snyder 1997, pp. 192–199), but the microfoundations behind many such conjectures about signaling effectiveness – as well as the

---

[2] The relative effectiveness of these measures, though, depends on the source of uncertainty: resolve or martial effectiveness (Arena n.d.). Signaling stories also presume that mobilization is public, but see Lai (2004) and Slantchev (2010) for the logic behind secret military preparations.

empirical conditions supporting them – remain unconnected to a theory that explicitly links intra-coalitional and crisis bargaining.

The chapter's key insight in this regard is that the political trades that states make to ensure military cooperation can often have a direct impact on the signaling value of their military threats and bargaining strategies. In Chapter 3's formal model, states bargain over shares of the crisis outcome, such as territorial spoils or indemnities, but in practice they also haggle intensely over bargaining and signaling strategies. As noted earlier, the United States balanced the desire to engage in large mobilizations against the need to ensure NATO partners' cooperation in crises over Berlin and Kosovo, but these crises are hardly unique. American restraint in the Korean War's early phases, which China in particular viewed as a signal of the expansiveness of American aims, was heavily influenced by the desire to maintain intra-coalitional cooperation (Stueck 1995, pp. 132–134), especially in the face of the British inability to commit to cooperating in Korea if the risk was touching off a war in Europe (p. 148).[3] Likewise, France found itself limited in the threats it could make toward Germany in early twentieth century crises over Morocco in 1905 and 1911; its chief ally, Russia, was unable to contribute militarily for all but the most serious stakes as it recovered from its defeat in the Russo-Japanese War in 1905 (see Stevenson 1997). Finally, the coalition arrayed against Qing China in the crisis over the Boxer Rebellion disagreed over the desirability of a naval demonstration aimed at signaling a willingness to impose its terms on the Empire by force (Xiang 2003, ch. 7).

The scope of a coalition's threats will generally be limited by the preferences of its least-resolute member, but when considered in the broader context of signaling and crisis bargaining, it is not clear that weak, diluted threats should always be problematic; as shown in the later discussion, the effect of moderated threats depends on what actions a lead state might have taken *absent* the need to ensure military cooperation. If a moderated threat masks a true willingness to fight, then the result – as anticipated by the extant literature (Christensen 2011, Lake 2010/2011) – is an otherwise avoidable war. However, if absent a coalition partner a lead state might have been tempted to bluff, to take steps that would generate a risk of war, then partners may discourage bluffing and preserve the peace. The challenge, then, is to identify the conditions under which coalition partners are likely to shape the course of international crises in each fashion. The theoretical model that follows addresses this problem by (a) linking divergent preferences over bargaining strategies to the uneven distribution of the costs of war within the coalition and (b) making a direct comparison to singletons facing the same signaling challenges.

---

[3] Pressures in Korea went both ways, however; the United States convinced South Korean leader Syngman Rhee to accept limits on war aims in return for more support for his government after the war (Stueck 2004, ch. 5).

4.2 A THEORY OF COALITIONAL CRISIS BARGAINING

Chapter 3 showed that lead states may take on partners with diverse preferences, despite a higher price of cooperation, in order to bolster their military prospects against their targets in the second stage of military multilateralism. However, it treated the crisis and some specifics of intra-coalitional bargaining in reduced form. To generate more precise expectations over how coalitional preferences and military power affect crisis bargaining, this chapter's theoretical model expands on the previous one by (a) introducing a crisis bargaining game under incomplete information and (b) modeling explicitly one issue over which coalition partners bargain quite often: military threats and mobilization strategies. Specifically, a lead state chooses a level of military mobilization, mindful of both whether an erstwhile coalition partner will find the attendant costs acceptable in the event of war and how it alters the target's assessment of the lead state's resolve.

Though more complicated, the model retains the basic logic of coalition formation, in that a partner with divergent preferences must be compensated for its cooperation. Specifically, coalition partners may have divergent preferences over mobilization levels because a variety of factors, from geography to resource constraints to domestic politics, lead to the perceived costs of war falling unevenly across its participants. For example, in the early phases of NATO's attempt to stop Serb ethnic cleansing in Kosovo, the United States and the United Kingdom favored strong escalatory steps in the form of legally binding mobilization and the forward deployment of aircraft, yet they faced objections from other member states worried over selling a costly ground war to their constituents, externalities from proximity to the war zone, or both (see Clark 2001, chs. 6, 7). In fact, of the states that contributed most to Operation Allied Force, public opposition to the possibility of a NATO-led war was greatest in the two countries nearest the Balkans: Germany and Italy (Auerswald 2004, pp. 640–641). As such, escalatory steps and public threats were dialed back, because American decision makers viewed these disagreements as potentially threatening to military cooperation in the event of war, as well as the future of the alliance (Auerswald 2004, Clark 2001). More generally, coalition partners may differ in their preferred levels of mobilization due to domestic constraints, exposure to the physical costs of war, or even their relative valuation of the issues at stake. As a result, they are likely to disagree over signaling and mobilization strategies, creating for coalition leaders a trade-off between substantial military mobilizations and the cooperation of their most cost-sensitive potential partners.

In the model that follows, a lead state and a potential partner both desire concessions from a target state, although they may differ in their sensitivity to the costs of war; specifically, the partner's costs for war increase in the lead state's chosen level of military mobilization. The target begins the game unsure of the leader's resolve, or its valuation of the stakes, rendering it uncertain over just how little it can concede and still avoid a costly war. The lead state has the opportunity to make a military threat – mobilization, in this case – that is costly up front, affects the military balance, and influences the partner's costs of fighting. Should bargaining break down in war, however, the partner may choose not to participate rather than expose itself to the costs of cooperation, creating a difficult trade-off for the lead state: mobilization may help signal its resolve, but if high mobilization also threatens to make the war too costly or destructive, it may come with the added cost of foregoing the partner's cooperation. Thus, to the extent that coalitions emerge in the model, they do so endogenously as a function of the partner's willingness to cooperate and the lead state's comparison of the benefits of unilateral and multilateral action. By integrating both coalition formation and crisis bargaining in the same model, I also ensure that my predictions are based on a selection process, which reduces the chances that selection bias plagues the results when I model the probability of war in a sample of observed crises later in the chapter.

4.2.1 The Model

Suppose that two states, a lead state (L) and target (T), are engaged in a dispute, in which L may act as a singleton or with a partner (P) to bolster its military prospects. Crucially, P may differ from the lead state in its assessment of the costs of war or its valuation of the stakes of the crisis.[4] For its part, the target begins the game unsure of the leader's valuation of the stakes, or resolve, rendering it uncertain over just how little it can concede and still avoid war; in this way, the game mirrors standard dyadic treatments of crisis bargaining under asymmetric information (e.g., Fearon 1995). Before the target makes a proposal, the lead state has the opportunity to make a military threat, such as mobilization, that is costly but also improves its military prospects (cf. Slantchev 2005, 2011). However, should bargaining break down in war, L's potential partner then chooses whether or not to participate in the conflict, creating for L a possible tradeoff: mobilization may demonstrate resolve, but if it also threatens to make cooperation unattractive for P, solving the target's information problem may come with the added cost of creating a commitment problem for L's potential coalition partner. Thus, the lead state's mobilization decision has implications for both signaling and coalition-building, which also ensures that, to the extent that coalitions emerge in equilibrium, they do so as a function of the partner's willingness to fight and the lead state's comparison of the benefits of ensuring and foregoing military cooperation.

---

[4] Consistent with the data collection effort and Chapter 3's theoretical model, L and T are thus the principal belligerents in the crisis, hence the use of the terms "lead state" and "partner" on L's side. However, these terms do not indicate any particular power relationship between lead state and partner – only that L will bargain with the target regardless of whether it receives P's support.

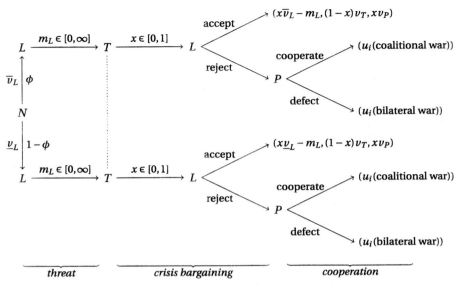

FIGURE 4.1. Signaling, crisis bargaining, and military cooperation

As shown in Figure 4.1, the game begins when Nature chooses L's valuation of the stakes under dispute, $v_L > 0$, such that the lead state is resolute ($v_L = \overline{v}_L$, placing a high value on the stakes) with probability ϕ and irresolute ($v_L = \underline{v}_L$, placing a lower value on the stakes) with probability $1 - \phi$, where $0 < \underline{v}_L < \overline{v}_L$. Nature reveals L's type only to L and P, although T knows the distribution from which it is drawn; thus, T's prior belief that the lead state is resolute is ϕ. P's and T's common knowledge valuations are $v_P > 0$ and $v_T = 1$, respectively. Note that P shares L's most preferred outcome, that is, full concessions from the target, but P may weigh the stakes either less ($v_P < v_L$) or weakly more ($v_P \geq v_L$) relative to the potential costs of war than the lead state. P also receives a payoff for the final distribution of the benefits regardless of whether it participates in a war. Thus, P has preferences over the issue, such as ethnic cleansing in Kosovo in the late 1990s, which a number of European states wished to see end regardless of whether they contributed to the military effort to compel the withdrawal of Serbian forces. In terms of the theoretical model in Chapter 3, the distributive or substantive outcome of the crisis is the public component of the partner's payoffs (analogous to γ), while the private component is analogous to its costs for participating in the war ($1 - \gamma$).

Next, L chooses a level of costly military mobilization, $m_L \geq 0$. In the restricted version of the model analyzed here, only two mobilization options exist, high ($\overline{m}_L$) and low ($\underline{m}_L$) where $0 < \underline{m}_L < \overline{m}_L$. These levels correspond to each type's optimal mobilization level, $\underline{m}_L$ for $\underline{v}_L$ and $\overline{m}_L$ for $\overline{v}_L$, given the military capabilities and strategies of the other players. I do this to simplify the analysis, trading a set of richer mobilization options for a more detailed treatment of bargaining dynamics.[5] The level of mobilization affects the lead state's payoffs in two ways. First, it sinks costs up front, such that L pays $-m_L$ directly. Second, it also contributes to the lead state's probability of victory in war by increasing its share of the total level of armaments in the crisis. In bilateral and coalitional wars, respectively, the lead state's side wins with probability

$$p_L(m_L, m_T) \equiv \frac{m_L}{m_L + m_T} \quad \text{and} \quad p_L(m_L, m_P, m_T) \equiv \frac{m_L + m_P}{m_L + m_P + m_T},$$

such that its prospects increase in its own mobilization, decrease in the target's military capabilities $m_T > 0$, and increase in the partner's military capabilities $m_P \geq 0$, if present. L captures the whole prize, v_L, if it wins, but it receives zero, yielding the whole pie to the target state, in defeat.

With the effects of mobilization defined, we can characterize the lead state's expected utility for the game if it ends in a coalitional war as

$$EU_L(\text{coalitional war}) = -m_L + \left(\frac{m_L + m_P}{m_L + m_P + m_T} \right) v_L - c_L, \qquad (4.1)$$

where $c_L > 0$ are L's costs for war. Note that a bilateral war – that is, one without a coalition partner – would involve different total armaments, or $(m_L + m_T)$, such that

$$EU_L(\text{bilateral war}) = -m_L + \left(\frac{m_L}{m_L + m_T} \right) v_L - c_L. \qquad (4.2)$$

Equations (4.1) and (4.2) show that mobilization plays a role similar to that in Slantchev's (2005) dyadic model of military threats, both sinking costs and, by increasing the attractiveness of military action, tying hands. However, as discussed in the equilibrium analysis that follows, mobilization in this model may also affect the strategies of both target states *and* potential coalition partners.

After observing L's mobilization level, which I equate with the size of its threat, the target then proposes a share of the stakes, $x \in [0, 1]$, to yield, and $1 - x$ to keep for itself.[6] If the lead state accepts, then the game ends with the target keeping $1 - x$, the lead state receiving $xv_L - m_L$ (the terms of settlement

---

[5] Technically, this simplifies the analysis of out-of-equilibrium beliefs and actions, which may be infinitely many when players can make small deviations from their optimal mobilization decisions that still affect the military balance and, as a result, the target's optimal proposal.

[6] The model abstracts away from any additional political costs that the target pays for making a generous offer to L after seeing a signal of resolve, which Huth (1988) argues must often come with a face-saving way for the target to climb back down the escalatory ladder. In this model, such costs would encourage T to risk war in cases where it otherwise might not; however, as long as such costs are not so high as to preclude peace in the first place, the reason that war occurs (or does not occur) in the present theory would be the same.

less mobilization costs), and P receiving xv_P.[7] Should the lead state reject the target's proposal, a war ensues in which P chooses whether or not to participate. If the potential partner cooperates, it fights alongside the lead state, and if it defects, L fights the target state alone in a bilateral war. This can be thought of as a coalition either forming or failing to form, ensuring that equilibrium coalitions are in the interest of both lead state and partner to build; since strategies are commonly known, whether a coalition forms or not will be known by all players as soon as the lead state chooses a mobilization level, allowing us to view the lead state's mobilization choice as both a military threat *and* a decision over whether to form a coalition.

If P cooperates to produce a coalitional war, the lead state's payoffs are as defined in Equation (4.1), while the target and partner receive

$$EU_T(\text{coalitional war}) = \frac{m_T}{m_L + m_P + m_T} v_T - c_T$$

and

$$EU_P(\text{coalitional war}) = \frac{m_L + m_P}{m_L + m_P + m_T} v_P - c_P m_L,$$

respectively. The target's costs for war are $c_T > 0$, but the partner's costs for war are also increased by the leader's chosen level of mobilization, such that it pays $-c_P m_L$ where $c_P > 0$. As discussed earlier, P may face domestic-political, resource, or geographical constraints that make it more difficult to participate in wars that follow larger mobilizations. Specifically, larger mobilizations might make participation in the war domestically unpopular, place greater burdens on financial and logistical capacity, or simply generate greater war costs for P due to proximity to the conflict zone or specialized efforts. As a result, in addition to their valuation of the stakes, L and P also differ in their sensitivity to the costs and destruction of war.[8]

If P defects, leaving L to fight T in a bilateral war, the belligerents' payoffs are as given above but for the absence of m_P in the total level of armaments. For its part, P gets

$$EU_P(\text{bilateral war}) = \frac{m_L}{m_L + m_T} v_P,$$

where it receives its valuation of the stakes if the leader wins the war alone. Thus, withdrawing cooperation – that is, refusing to join the coalition – reduces the chances that L wins the war, but it also reduces P's exposure to the costs

---

[7] The model thus introduces explicit bargaining where both Slantchev (2005) and earlier models (Fearon 1997, Schultz 1998) assume that the stakes are indivisible.

[8] I have also solved a more complicated version of the model in which all players' costs for war are a function of the total level of armaments involved, or the scope of the war, such that a belligerent i pays costs $c_i(m_L + m_P + m_T)$. However, since the main substantive results are the same, I opt here for the simpler specification in which only P's costs for war directly increase in L's mobilization.

of fighting, allowing it to enjoy the public component of the crisis outcome but pay none of the private costs. I show in the next section how the lead state's desire to influence this decision, and T's ensuing choice over cooperation and defection, can endogenously influence both the probability of war and the formation of coalitions.

We can now characterize each player's payoff function over terminal nodes of the game, for any arbitrary combination of player-types v_L, mobilization levels m_L, and proposals x. The lead state's function is given by

$$u_L = \begin{cases} -m_L + x v_L & \text{if settlement} \\ -m_L + \left(\frac{m_L}{m_L+m_T}\right) v_L - c_L & \text{if bilateral war} \\ -m_L + \left(\frac{m_L+m_P}{m_L+m_P+m_T}\right) v_L - c_L & \text{if coalitional war,} \end{cases} \quad (4.3)$$

while the potential partner's is

$$u_P = \begin{cases} x v_P & \text{if settlement} \\ \left(\frac{m_L}{m_L+m_T}\right) v_P & \text{if bilateral war} \\ \left(\frac{m_L+m_P}{m_L+m_P+m_T}\right) v_P - c_P m_L & \text{if coalitional war,} \end{cases} \quad (4.4)$$

and the target's is

$$u_P = \begin{cases} (1-x) v_T & \text{if settlement} \\ \left(\frac{m_T}{m_L+m_T}\right) v_T - c_T & \text{if bilateral war} \\ \left(\frac{m_T}{m_L+m_P+m_T}\right) v_T - c_T & \text{if coalitional war.} \end{cases} \quad (4.5)$$

These utility functions capture the essentials of both (a) crisis bargaining under asymmetric information, as the target is unsure what bargains the lead state accepts, and (b) the transactional nature of coalition-building, as the lead state must choose the proper terms of compensation if it wishes to secure the cooperation of a partner that, all else being equal, would prefer to stay out of the crisis. In the language of the model of the previous chapter, the extent to which L's and P's preferences diverge can be captured in differences in either valuations of the stakes (v_L and v_P) and sensitivity to the costs of war (c_L and c_P), while the concession made to adjust for these differences emerges in the choice of mobilization level: the smaller L's mobilization, the more it concedes to its partner.

Finally, note that other signaling models involving multiple players generally identify conditions under which third parties discourage bluffing. In Smith's (1998) model, national leaders wish to signal competence to both international enemies and their domestic constituencies, with the result that only "low" types, or incompetent national leaders, make empty threats during crises. Similarly, Schultz (1998) shows that a domestic opposition can also discourage bluffing and enhance the credibility of threats. However, in neither case can the leader's second audience affect its military prospects, nor does the audience have preferences over the material content of the signal, as P does here. I show in the following section that, while P can indeed discourage irresolute or "low" types of lead state from bluffing about their willingness to fight, the material consequences of military threats as signals can also encourage *resolute* lead states to mask their resolve despite bilateral incentives to reveal it.

4.2.2 Analysis

How does a lead state's desire to secure and maintain military cooperation affect signaling behavior in crises? How does a partner's wish to restrain the lead state shape the chances of war? Finally, why and under what conditions do coalitions differ from singletons in the probability that their crises escalate to war? To answer these questions, I begin this section with a brief characterization of equilibria in a version of the game without a potential coalition partner to show that the underlying model produces typical signaling dynamics, which establishes a baseline case against which to compare equilibria involving the potential partner, then show how the addition of that potential coalition partner changes several important features of the crisis, including the probability that it escalates to war.

Signaling dynamics emerge in three different kinds of equilibrium, differentiated by the actions that L takes when it is of a particular type and what this mapping of types to actions reveals to the target state about L's resolve. First, in a *separating equilibrium*, L takes an action that is unique to its type – for example, the irresolute type mobilizing low and the resolute type mobilizing high – that reveals its type to the target. Separating equilibria are peaceful, because a unique mapping of types to actions solves the target's information problem and allows it to tailor its proposal so as to minimize concessions and still avoid war. Second, in a *pooling equilibrium*, L takes the same action regardless of its type, which provides no information to the target; as such, T can only act on the beliefs with which it begins the game, which may encourage it to make a proposal that L rejects, leading to war. Finally, in a *semi-separating equilibrium*, a player takes one action if it is of one type (mobilizing high if resolute) but randomizes over its actions if it is of another type (in this case, randomizing over high and low mobilizations if it is irresolute). Semi-separating equilibria also entail a chance of war, because if L mobilizes high, it may be either resolute or irresolute; thus, while T's uncertainty is diminished in this equilibrium, it may not learn enough to enable it to avoid war.

The two-player game entails both peaceful separating and violent semi-separating equilibria. The full three-player model, however, produces a contrasting set of results, which I analyze by drawing a distinction between two sets of conditions. First, I discuss equilibria with a committed partner, or one that cooperates in a war regardless of L's mobilization decision. Then, I consider the case in which P cooperates only if L chooses low mobilization – that is, P is "skittish" – which means that L must forfeit a coalition if it mobilizes enough to reveal its type to the target; with a skittish partner, the reliability of military cooperation, and thus the distribution of power, becomes an open question. While a committed partner produces equilibria isomorphic to the two-player case, two differences emerge in the model with a skittish partner. First, separating and semi-separating equilibria both exist, although separating equilibria are more common when the target is strong. Second, when the target is weaker, there exists a pooling equilibrium unique to the skittish case in which both types of L choose *low* mobilization, preserving both military cooperation and, tragically, the target's uncertainty, which entails a risk of war.

I assume throughout the analysis that the target's prior beliefs over the lead state's resolve are sufficiently optimistic – that is, confident that L is irresolute – that, in the absence of any information revelation, T is willing to risk war. This serves two purposes. First, if war is to occur *or* be averted, the target must be willing ex ante to risk it; if its prior beliefs do not support making proposals that risk war in the first place, then all equilibria are peaceful, and uninteresting, by default. Second, any peaceful equilibria that do exist will derive from L's ability to credibly signal its type rather than some characteristic of the strategic environment unrelated the tensions of intra-coalitional politics.

The Two-Player Game

The two-player version of the game has two equilibria. First, in the separating equilibrium, each player-type of L chooses a unique mobilization level, high if resolute and low if irresolute, that reveals its type and facilitates peaceful settlement. Second, in the semi-separating equilibrium, the resolute type chooses high mobilization, but the irresolute type probabilistically bluffs with high mobilization, hoping to convince the target to offer better terms by mimicking the behavior of the resolute type. As a result, T only partially updates its beliefs following high mobilization, which tempts it to probabilistically risk war in response, randomizing between a proposal sure to be accepted and one that L will reject if it turns out to be the resolute type.

Proposition 4.1 (Two Players). *In the two-player game, the following sets of strategies are each part of a Perfect Bayesian Equilibrium:*

(a.) *When $m_T \geq \hat{m}_T$, player-types take unique actions and there is no risk of war (separating equilibrium).*

(b.) *When $m_T < \hat{m}_T$, $\underline{v}_L$ bluffs probabilistically, generating a risk of war (semi-separating equilibrium).*

See appendix for proof.

In the separating equilibrium, each type of L chooses a level of mobilization that balances its upfront costs against improved military prospects and distinguishes – or separates – it from the other type. Since resolute and irresolute types value the outcome differently, they choose levels of mobilization unique to each type and from which neither type would wish to deviate, given T's anticipated response. As a result, T believes L to be resolute only if it mobilizes at the resolute type's optimum, $m^* = \overline{m}_L$ (high), and irresolute at the irresolute type's optimum, $m^* = \underline{m}_L$ (low). After observing L's mobilization, T proposes terms that reflect its beliefs, updated according to L's signal. The game thus ends peacefully, because T, knowing exactly what proposals will provoke war, can shape its aims efficiently and save the costs of war.

Proposition 4.1 states that the separating equilibrium exists when the target is sufficiently powerful, or $m_T \geq \hat{m}_T$, where $\hat{m}_T$ is the critical value that separates the separating and semi-separating equilibria. When $m_T \geq \hat{m}_T$, the lead state must mobilize so much to offset the target's strength that the irresolute type finds the costs of bluffing – that is, high mobilization – too expensive.[9] Rather than pay the costs of high mobilization to receive a better deal, the irresolute type is content to reveal itself as such through low mobilization. However, when the target is weaker, the difference between optimal low and high mobilization levels is smaller, and the irresolute L finds the cost of bluffing up to the level of high mobilization more tolerable. This link between target strength and the costs of bluffing through mobilization is critical to the strategic dynamics of both two- and three-player variants of the game.

In the semi-separating equilibrium that exists when the target is weaker, or $m_T < \hat{m}_T$, bluffing is not prohibitively expensive; the difference in mobilization levels that L finds optimal when resolute and irresolute is smaller, thanks to the reduced requirements for mobilizing against a weaker target. Thus, L's desire to misrepresent its type when it is irresolute is easier to satisfy. While the resolute type chooses high mobilization, the irresolute type randomizes between low and high mobilization, occasionally bluffing and refusing to reveal the extent of its irresolution in order to persuade T to offer better terms. This prevents T from updating its beliefs fully over L's type if it observes high mobilization, and its own mixing strategy dictates that it respond by occasionally making a proposal that will be rejected; otherwise, it would forego entirely the chance to secure the best possible terms against L. As a result, it sometimes makes a proposal that the irresolute type accepts but that the resolute type rejects, leading to war. Thus, in the semi-separating equilibrium, L's incentive to misrepresent its resolve when it is truly irresolute generates a risk of war, because the costs of high mobilization cannot deter it from mimicking the resolute type as it is in the separating equilibrium.[10]

---

[9] To see this formally, note that the difference between these two mobilization levels increases in m_T, or $\partial \left(\overline{m}_L - \underline{m}_L \right) / \partial m_T > 0$.

[10] Mixed strategies like those found in the semi-separating equilibrium are designed to blunt the advantage another player might gain by perfectly anticipating one's actions. Therefore, both

The two-player model reproduces the standard story of signaling and crisis bargaining: when a resolute type can take an action too costly for an irresolute type to mimic, peace is possible through costly signaling. When bluffing is less expensive, however, an irresolute type may bluff, behaving in such a way as to preserve the target's uncertainty and generate a risk of war. Doing so allows it to increase the chances of receiving better proposals, but the cost is a risk of war – a cost paid by the resolute type, because it cannot distinguish itself from an irresolute type mimicking it. However, the Berlin and Kosovo cases described earlier involve not irresolute types trying to appear resolute, but ostensibly resolute states taking steps that would *not* demonstrate resolve, limiting mobilization because of what appears to be divergent preferences within a coalition. When will coalitions partners exert such downward pressure, why do lead states tolerate it, and what are the implications for the probability of war?

The Three-Player Game

How do coalition politics and the desire to secure military cooperation alter the standard account of costly signaling? I show in this section that introducing a potential coalition partner into the game can either increase *or* decrease the probability of war, depending on the partner's preferences and the target's military strength. I frame the discussion around P's choice over cooperating or defecting should bargaining break down in war, particularly the conditions under which P cooperates regardless of L's mobilization decision (committed) or cooperates only if the leader chooses low mobilization (skittish). Giving the potential partner this commitment problem – and letting the lead state anticipate it – ensures that only those coalitions that both players find attractive form in equilibrium. Thus, comparisons across two- and three-player games are not made based on coalitions that neither player would have an interest in forming, meaning that the lead state's mobilization decision can also be understood as an endogenous choice over coalitional or unilateral action, integrating the partner selection process from the first theoretical model into the second and ensuring that issues of strategic selection are taken into account in the empirical implications derived from the model.

Faced with the choice of fighting alongside the lead state or defecting in the game's final move, P plays one of three strategies: (a) cooperating regardless of L's mobilization choice, (b) cooperating only in the event of low mobilization, or (c) defecting regardless of mobilization levels. Since coalitions never form in the last case, I focus here on the first two, where coalitions can – but need not – form in equilibrium. In the case of a committed partner, described in Proposition 4.2, P's costs for war are sufficiently low, or $c_P < c_P^l$, that it cooperates regardless of L's mobilization decision. In other words, the lead

players randomize; each ensures that it avoids the worst outcome in expectation while accepting that the other will prevent it from reaching its own best outcome; technically, players render one another indifferent over available actions, freeing them to randomize between them.

state can choose its mobilization level safe in the knowledge that its partner will ultimately cooperate in the war, because P's exposure to its costs is sufficiently limited.

Proposition 4.2 (Three Players, Committed Partner). *When P is committed, or $c_P < c_P^l$, the following sets of strategies are each part of a Perfect Bayesian Equilibrium:*

(a.) *When $m_T \geq \hat{m}_T$, player-types take unique actions and there is no risk of war (separating equilibrium).*

(b.) *When $m_T < \hat{m}_T$, $\underline{v}_L$ bluffs probabilistically, generating a risk of war (semi-separating equilibrium).*

See appendix for proof.

When P is *committed*, it cannot credibly threaten to defect in the event of war even if the lead state chooses high mobilization. Therefore, even though it prefers low mobilization to high, all else being equal, P imposes no constraints on either an irresolute lead state's incentive to bluff or a resolute lead state's willingness to reveal itself through high mobilization; in other words, P is an inexpensive, easily compensated partner. As a result, the same two equilibria, separating and semi-separating, exist in the case of a committed partner as in the two-player case. In fact, since the lead state can economize on mobilization costs thanks to its partner's unconditional support, its mobilization levels take its improved military prospects into account, with the result that the same threshold found in the two-player game, $\hat{m}_T$, divides separating from semi-separating equilibria in this case. Thus, easily constructed coalitions built around expectations of unconditional cooperation should be just as prone to probabilistic bluffing and potential war as states that do not take on coalition partners, because any threats of the partner's subsequent defection are incredible.

On the other hand, when P is *skittish*, it conditions its cooperation on L's mobilization choice, cooperating if $m_L = \underline{m}_L$ (low mobilization) and defecting if $m_L = \overline{m}_L$ (high mobilization). In this case, the partner can credibly promise to fight alongside the lead state only if L chooses low mobilization, because high mobilization imposes intolerably high political or material costs. Whatever statements P might make at the outset of the game, the fact of moving last ensures that it cannot commit to tolerating high mobilization, and P follows this strategy when $c_P^l \leq c_P < c_P^h$, or when it is neither too sensitive nor too insensitive to the costs of war. Put differently, high mobilization entails costs too large, $c_P \overline{m}_L$, for P to justify cooperation; instead, it defects and saves the costs of war, leaving L to fight on as a singleton. On the other hand, low mobilization is less costly, since $c_P \underline{m}_L < c_P \overline{m}_L$, rendering cooperation in a coalitional war tolerable.[11] Thus, in contrast to the case of a committed partner, a skittish P may be able to use its threat to refuse cooperation to moderate L's bargaining position – seen in American concerns over the commitment of European allies over Berlin and Kosovo – but as I show next, restraining L in this fashion can have unexpected consequences for the probability of war.

When P conditions its cooperation on mobilization levels, the lead state faces a stark trade-off. In the absence of a skittish partner, high mobilization carries only the direct costs of mobilizing, but a skittish partner introduces an additional, indirect cost: a foregone improvement in the military balance resulting from P's refusal to cooperate. Thus, high mobilization may demonstrate resolve, but it carries the opportunity cost of fracturing a potential coalition if bargaining breaks down in war, such that L can credibly threaten only to fight alone. High mobilization, then, amounts to an explicit choice to pay these opportunity costs – that is, to act unilaterally. Low mobilization, on the other hand, ensures the partner's cooperation as part of a potential coalitional war but fails to convince the target of L's resolve, tempting T to risk war with its proposal. How does the lead state manage this tension? As it does in Propositions 4.1 and 4.2, the answer turns on the target's military capabilities.

Proposition 4.3 (Three Players, Skittish Partner). *When P is skittish, or $c_P^l \leq c_P < c_P^h$, the following sets of strategies are each part of a Perfect Bayesian Equilibrium:*

(a.) *When $m_T \geq \max\{m_T^{\dagger}, \tilde{m}_T\}$, player-types take unique actions and there is no risk of war (separating equilibrium).*

(b.) *When $m_T^{\dagger} \leq m_T < \tilde{m}_T$, $\underline{v}_L$ bluffs probabilistically, generating a risk of war (semi-separating equilibrium).*

(c.) *When $m_T < m_T^{\dagger}$, both player-types choose low mobilization, generating a risk of war (pooling equilibrium).*

See appendix for proof.

Proposition 4.3 states that separating and semi-separating equilibria exist in the skittish partner case, as they do with two players and a committed partner. However, an additional equilibrium is unique to the skittish partner case: a pooling equilibrium in which both types of L choose low mobilization. In other words, and contrary to the expectations of signaling models in which separation is the chief goal (e.g. Fearon 1997, Slantchev 2011, Smith 1998), the resolute type mimics the irresolute type. Low mobilization ensures the skittish partner's cooperation in the event of war, but the weak threat required to build the coalition also prevents a resolute lead state from revealing its type, and the

---

[11] It is worth noting that P can be skittish due to its aversion to high mobilization *or* a low valuation of the stakes. In formal terms, P is also skittish for middling values of v_P, or $v_P^l < v_P \leq v_P^h$, but I focus on P's direct aversion to the costs of mobilization to simplify the discussion of the results.

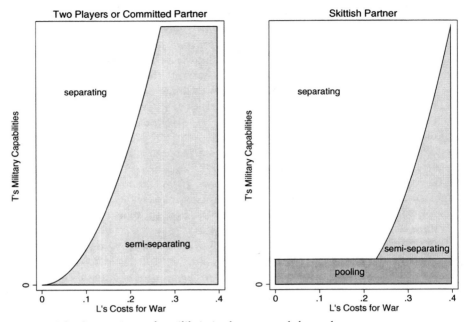

FIGURE 4.2. Comparison of equilibria in the two- and three-player games

target's information problem remains unsolved. The lead state's threat does not alter the target's beliefs, and when the target is sufficiently optimistic that the lead state is irresolute, it balances the risk of war against a miserly proposal that only the irresolute type will accept. Thus, the desire to preserve military cooperation leads a resolute lead state to engage in behavior that generates a risk of war – that is, refusing to reveal its private information.

How do these equilibria compare to those of the two-player and committed partner cases? Figure 4.2 plots the equilibrium spaces, on identical scales, for each case as a function of two exogenous variables: the leader's costs for war, c_L, and the target's military strength, m_T. In the two-player and committed partner cases, represented in the left panel, a single threshold value, $\hat{m}_T$, divides separating and semi-separating equilibria, and the risk of war is indicated by the shaded area under the curve. In the three-player game with a skittish partner, presented in the right panel, a similar threshold, $\tilde{m}_T$, divides separating from semi-separating equilibria. However, it falls farther to the right, and the separating equilibrium exists over a wider range of the parameter space than it does at left – but only as long as the target is sufficiently powerful, or $m_T \geq m_T^{\dagger}$. Thus, the presence of a skittish partner with a strategy of conditional cooperation can *reduce* the probability of war by discouraging the irresolute type from bluffing – that is, by restricting the conditions under which risky semi-separating equilibria exist. In the separating equilibrium, the threat of losing P's military support and the attendant opportunity costs outweigh the benefits of bluffing, and the irresolute type is happy to reveal itself as such when it receives coalitional support in return. In this case, and quite contrary to the expectations of most work focused on such "watered-down" threats, the partner's attempt to moderate the lead state's bargaining position effectively discourages wars that, without the possibility of a coalition, would otherwise occur.

While the need to ensure a skittish partner's cooperation can discourage risky bluffing, it also enables resolute leaders to use unilateral action to credibly signal their type by paying the opportunity costs of foregone military cooperation. In the two-player game, or in the presence of a committed partner, the resolute type is unable to separate itself when $m_T < \hat{m}_T$, because the costs of bluffing are low; however, semi-separating equilibria in which the irresolute type bluffs occupy a much smaller part of the parameter space when P is skittish, that is, where $m_T^\dagger \leq m_T < \tilde{m}_T$. This is unsurprising, as bluffing is doubly costly when L must include P in its calculations; not only must the irresolute type mobilize more than it wishes but it must also pay the opportunity costs of losing the capabilities of a coalition partner. For its part, the resolute lead state is happy to forego coalition building in order to mobilize at its optimal level, reveal its type, and receive a favorable bargain at no risk of war. Thus, for resolute types to form coalitions with skittish partners, they must be willing to tolerate low mobilizations, which is easiest when low and high mobilization levels are not too different – that is, when the target is relatively weak, which ensures that the upfront costs of bluffing are not too high.

This separating equilibrium is intriguing because it indicates that states known to have available coalition partners can forego securing their cooperation, "going it alone" so to speak, and use those opportunity costs to reveal their true level of resolve. In short, the very choice of acting unilaterally may serve as a costly – and therefore credible – signal of resolve. Many mechanisms proposed to increase the credibility of threats involve the direct costs of arming (Fearon 1997) and potential domestic punishment for backing down from public threats (Fearon 1994) but foregoing the chance to build a coalition, when one's enemy knows such a coalition was possible, might offer an additional, internationally based mechanism for signaling resolve. The larger the opportunity costs of acting without partners, the more easily resolute states can signal their resolve by doing just that – acting without them.[12] In this case, third parties influence the probability of war even if they make no threat to get involved, suggesting another factor, in addition to the measurement of power (Croco and

---

[12] Acting unilaterally can only serve as a signal, of course, when the underlying bargaining problems is rooted in private information about resolve. If, for example, the issue at the root of the crisis is a commitment problem, as it was in crisis that led to the Iraq War of 2003, then unilateral action's signaling value is moot.

Teo 2005, Sobek and Clare 2013), that makes dyadic studies problematic for the study of coalitions.

However, as shown in Figure 4.2, the increased prevalence of separating, and therefore peaceful, equilibria occurs *only* against relatively strong targets. When the target is sufficiently weak in terms of military capabilities, or when $m_T < m_T^\dagger$, the lead state chooses low mobilization regardless of its type, preserving the target's uncertainty and generating a risk of war that would not otherwise exist were types to separate. Specifically, the resolute type economizes on mobilization against a weak target, preserving cooperation and bringing P's capabilities to bear in the crisis, even when doing so tempts the target to make a risky proposal that the lead state is sure to reject if it is the resolute type.[13] Thus, the resolute type of L opts for a coalitional war rather than high mobilization and a credible unilateral signal. Notably, the possibility of war emerges from a partner's sincere desire to restrain the lead state's bargaining position, hoping to protect itself against the problems of moral hazard that emerge for committed partners, particularly when high mobilizations will saddle it with a disproportionate share of the costs of war. Nonetheless, the consequence of moderating L's bargaining posture can have the tragic effect of raising the chances of war; indeed, the very attractiveness of cooperation creates a risk of war where, absent the chance to build a coalition, types would reveal themselves peacefully. Finally, the requirement that the partner not be too powerful, or $m_T < m_T^\dagger$, implies that a potential partner's effect on the probability of war is conditional not only on its own preferences but also on the military strength of the target.

The logic supporting the pooling equilibrium is consistent with the notion that coalition partners can dilute or weaken a coalition's threats; in fact, coalitions only form with skittish parners around the less serious threats implied by low mobilization. As discussed at gerater length later in this chapter, it also provides a plausible account of the limited scope of NATO's threats leading up the the 1999 Kosovo War, thanks to several important but skittish partners and a militarily weak target state. In fact, Papayoanou (1997) argues that moderation can also explain NATO's Bosnia policy from 1991–1995, and the present model generalizes this insight by showing that the weakness of the target can facilitate both coalition formation and, by discouraging resolute states from revealing their type through high mobilization, the escalation of crises to war.

If pooling equilibria exist when targets are relatively weak due to the reduced costs of bluffing through mobilization, then the relationship between skittish potential partners and the probability of war – particularly as compared to singletons – is a conditional one.

---

[13] Though their mechanisms involve catching opponents unprepared for war, Slantchev (2010) and Trager (2010) also derive results where strong or resolute states mask their type and increase the chances of war by foregoing the chance to reveal private information.

Proposition 4.4. *When P is skittish and $m_T \geq m_T^\dagger$, a coalition's probability of war is higher than a singleton's, but when P is skittish and $m_T < m_T^\dagger$, a coalition's probability of war is lower than a singleton's. See appendix for proof.*

Notably, extant accounts of the effect of coalition partners on the chances of successful coercion do not take characteristics of the target into account (e.g., Bellamy 2000, Byman and Waxman 2002, Christensen 2011, Lake 2010/2011), focusing almost solely on divergent interests within coalitions. For example, Lake (2010/2011) asserts that "multiple actors appear to magnify problems of asymmetric information and costly signaling" (p. 9), but Figure 4.2 shows that this need not be the case. In fact, the presence of a third player – in this case, a skittish partner that can threaten to withdraw military support if it will find the war too costly – can *mitigate* signaling problems by discouraging an irresolute type from bluffing, which would otherwise generate a risk of war if L were to act as a singleton.[14] Thus, the partner's threat to withdraw cooperation forces a moderation of the lead state's bargaining position, and when it makes bluffing unattractive, that moderation facilitates a peaceful settlement.

On the other hand, when the target is not too powerful, or $m_T < m_T^\dagger$, the presence of a partner *does* increase the chances of war, precisely because of P's credible threat to defect. The partner's desire to limit the lead state's mobilization is successful, but the steps L must take to ensure cooperation have the perverse effect of rendering this less costly war more likely to happen. Without a skittish partner, war does not occur against a weak target when $\hat{m}_T \leq m_T < m_T^\dagger$, and it occurs probabilistically when $m_T < \min\{\hat{m}_T, m_T^\dagger\}$. With a skittish partner and a weak target, however, war occurs where it otherwise would not above $\hat{m}_T$ (left of the left panel's curve), and T is sure to risk war below $\hat{m}_T$ (right of the left panel's curve). Proposition 4.5 summarizes this relationship between member preferences in coalitions, target strength, and the probability of war.

Proposition 4.5. *Coalitions involving skittish partners are weakly more likely to go to war against weaker targets than against strong targets. See appendix for proof.*

Put differently, the moderating effects of skittish partners make war likelier against weaker targets than it is against stronger targets. Crises involving coalitions may thus escalate to war at more skewed distributions of power than bilateral crises, not because increasing relative power drives up their demands,

---

[14] Favretto (2009) also shows that third parties can facilitate peace under incomplete information by contributing observables that outweigh the impact of unobservables, making uncertainty less relevant rather than altering signaling and mobilization strategies.

but because it encourages resolute types to economize on mobilization, masking their resolve and tempting the target state to make a miserly offer of concessions that risks war. On the other hand, coalitions with inexpensive, credibly committed partners – those who cannot credibly threaten to defect – should have little effect on the relationship between power and war relative to singletons because they are unable to play the same role in the signaling process as skittish partners.

The model also sheds light on the conditions under which a skittish partner exercises the most downward pressure on mobilization. The threshold of target strength below which the lead state chooses low mobilization regardless of type is $m_T < m_T^\dagger$, where

$$m_T^\dagger = \frac{m_P^2 \underline{v}_L}{\left(\sqrt{\overline{v}_L} - \sqrt{\underline{v}_L}\right)^4}. \tag{4.6}$$

This threshold rises, becoming easier to satisfy, as the partner's military contribution, m_P, increases; the more powerful a skittish partner, the more L can economize on mobilization costs by retaining P's support, and L will choose low mobilization against ever more powerful targets. More powerful partners can thus exercise greater leverage over the coalition's bargaining position, which is unsurprising. However, it also implies that militarily powerful, and otherwise attractive, partners can *raise* the probability of war when they are skittish, because leaders are increasingly willing to mask their resolve, choosing low mobilization in order to retain valuable military support. We can draw a contrast here with the alliance literature (Leeds 2003b, Morrow 1994, Smith 1995), where more powerful allies should, all else being equal, reduce the probability of war by diminishing the role of unobservable factors in the crisis. In the present theory, however, the partner's ability to shift the terms of cooperation in its favor, to make fighting the war attractive, raises the probability of war when the lead state is willing to meet those terms. This particularly tragic outcome is increasingly likely when the skittish partner in question is also militarily valuable.

Proposition 4.6. *The more powerful a skittish partner, the more willing is the leader to send a weak signal that entails a risk of war. See appendix for proof.*

However, it is important to note that, while increasingly powerful skittish partners make resolute types more willing to mask their resolve, they also lower the threshold below which irresolute types probabilistically bluff, $\tilde{m}_T$, as part of the semi-separating equilibrium. Thus, while increasing partner strength makes one particular path to war – pooling on low mobilization – more likely, its total effect on the probability of war is ambiguous. Therefore, powerful partners should be associated with both stronger incentives for resolute lead states to bluff *and* stronger incentives for irresolute states *not* to bluff, such that there should be no consistent, bivariate relationship between partner strength and the probability of war.

Summary of Empirical Implications

Skittish coalition partners pose a unique challenge to lead states trying to use mobilization to demonstrate resolve because they create a commitment problem within the coalition. Against powerful targets, irresolute lead states prioritize maintaining cooperation, saving the upfront costs of bluffing and minimizing the risk of war. However, against weaker targets, resolute lead states mask their resolve, mobilizing too little to send a credible signal, ensuring a partner's cooperation at the cost of creating a risk of war – a risk that would not exist if cooperation were not possible. Thus, the effects of "weak" threats on the chances of war depend on *why* such threats are made; a partner's ability to limit the leader's mobilization is a force for peace at some times, but it can encourage war at others. The model also helps delineate the empirical conditions under which coalition leaders may rationally choose either to mask their resolve, thereby increasing the chances of war, or to honestly reveal their type despite ostensible incentives to bluff. Notably, these effects are conditional on the military strength of the target. Finally, it suggests that overcoming signaling difficulties may involve not only political features of the state in question (cf. Schultz 1998, Smith 1998) but also intra-coalitional politics. In fact, where some work argues that the presence of multiple actors in a crisis setting exacerbates the information problems that may cause war (Christensen 2011, Lake 2010/2011), introducing a possible coalition partner into a crisis environment can increase *or* decrease the probability of war, depending on the relative capabilities and preferences of coalition members and their enemies. Finally, foregoing a coalition partner that a target knows to be available can itself serve as a signal of resolve.

4.3 AN EMPIRICAL MODEL OF CRISIS ESCALATION

How do coalition partners affect the escalation of crises to war, and how do coalitions' rates of escalation compare to singletons? Chapter 2 showed that coalitional crises escalate to war more often than those involving singletons, but what factors drive this relationship? Further, is it true of all coalitions, or is it a conditional relationship, as indicated by the theoretical model? Scholarship is generally skeptical of coalitions' ability to credibly signal their resolve to fight (Christensen 2011, Lake 2010/2011), but the theory presented in this chapter suggests that this need not be the case. Specifically, coalitions should see higher probabilities of war than singletons should when their targets are sufficiently weak, but the probability of war should be comparatively *lower* when their targets are strong. On the other hand, there should be a much weaker relationship between the target's military strength and the probability that the crisis escalates to war in the absence of a coalition. In this section, I use the data first presented in Chapter 2 to test these hypotheses against the record of coalition formation and crisis escalation from 1946–2001.

4.3.1 Hypotheses

Before describing the research design, I use this section to translate the logic of the theoretical model into explicit hypotheses over the probability that crises escalate to war. While Figure 4.2 shows that there is an equilibrium relationship between coalition partners, target capabilities, and crisis escalation, it only indicates those regions of the parameter space in which war is possible rather than outline precisely how the probability of war changes as a function of the parameters. Therefore, I derive the equilibrium probability of war as a function of (a) the presence/absence of a skittish coalition partner and (b) the target's military strength, plot the expected relationship graphically, and state the hypotheses in terms of the empirical model estimated later in the chapter.

Translating any theoretical model into an empirical one requires a number of assumptions (Primo and Clarke 2012). I make an important one here by drawing a distinction between only bilateral and coalitional crises – not between types of partner in particular coalitions, whether skittish or committed. This is due to the difficulties of measuring preferences over specific escalation strategies (skittishness) in individual crises, for which the revealed-preference data (Reed et al. 2008, Strezhnev and Voeten 2013) used in Chapters 3 and 5 are ill-suited. However, there is little reason to think that it should cause problems for the analysis. Recall that equilibria in the two-player and skittish partner cases are isomorphic, such that, if it were possible to identify committed partners, they should display a relationship between target strength and war more similar to that in singletons' crises than crises involving coalitions with skittish partners. Therefore, since states acting unilaterally by definition have no skittish partners, the inclusion of committed partners alongside skittish ones into the category of coalitions should, at worst, produce aggregation bias, skewing results in favor of finding no difference between the presence or absence of coalition partners. As a result, the empirical tests in the following section are conservative, which should increase our confidence in any results showing differences across singletons' and coalitions' crises.

Beginning with the case of a bilateral crisis, where L is a singleton, Proposition 4.1 states that the probability of war declines in the target's military strength. Specifically, when $m_T < \hat{m}_T$, the irresolute L randomizes between low and high mobilization, while the target randomizes between risking war and buying off both types in the event of high mobilization; otherwise, when $m_T \geq \hat{m}_T$, the probability of war is zero. In the semi-separating equilibrium that exists for $m_T < \hat{m}_T$, war occurs in equilibrium at the confluence of three probabilistic events: (a) an irresolute lead state bluffs with high mobilization (h_2^*), (b) the target makes a proposal that risks war (r_2^*), and (c) the lead state

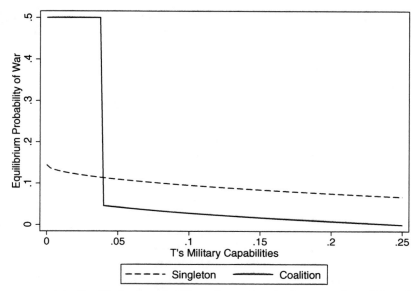

FIGURE 4.3. Equilibrium probability of war for singletons and coalitions by target military capabilities

turns out to be the resolute type (ϕ).[15] Thus, the equilibrium probability of war is $\Pr(\text{War}) = h_2^* r_2^* \phi$, which, as shown by the dashed line in Figure 4.3, decreases – albeit slowly – in the target's strength.[16]

On the other hand, in coalitional crises with a skittish partner, the function defining the equilibrium probability of war is slightly more complicated. As stated in Proposition 4.3, the probability of war is again zero when the target is sufficiently strong, or $m_T \geq \max\{\tilde{m}_T, m_T^{\dagger}\}$, but the semi-separating equilibrium exists only for middling values of target strength, or $m_T^{\dagger} \leq m_T < \tilde{m}_T$; in the three-player game, the target is sure to risk war when it is weak, or when $m_T < m_T^{\dagger}$. Thus, when $m_T^{\dagger} \leq m_T < \tilde{m}_T$, the probability of war is $\Pr(\text{War}) = h_3^* r_3^* \phi$, while the probability of war for weak targets is simply ϕ, or the probability that the lead state turns out to be strong, given that the target has risked war. As shown by the solid line in Figure 4.3, this implies a probability of war than begins substantially higher than in the singleton case before decreasing rapidly, falling below the singleton's probability of war, as the target's military strength increases.

---

[15] Note that, while in the formal model the analyst knows L's type, this is not the case in the empirical model, so the analyst shares T's probabilistic beliefs about the lead state's type, ϕ.

[16] Parameters are fixed at $\underline{v}_L = 0.25$, $\bar{v}_L = 1.25$, $c_T = 0.05$, $c_L = 0.4$ (ensuring skittishness), and $\phi = 0.5$ (ensuring that T will risk war under its priors).

From these graphs, we can derive two hypotheses over the relationship between coalitions, target strength, and the probability of war that I will test below.

Hypothesis 4.1. *When side 1 is a singleton, the probability of war decreases slowly in the strength of the target.*

Hypothesis 4.2. *When side 1 is a coalition, the probability of war decreases rapidly in the strength of the target.*

I turn next to a discussion of the research design necessary to uncover these relationships in the coalitions data described in Chapter 2.

4.3.2 Research Design

To test Hypotheses 4.1 and 4.2, I once again leverage the data on coalition participation in international crises first presented in Chapter 2, though the data now require a different format. Where Chapter 3 examined dyads of lead state and potential partners on each side in every crisis, the unit of observation in this chapter is the directed crisis-dyad, where side 1 may be either a coalition or a singleton. In this section I discuss the construction of the dataset, relevant theoretical variables, the choice of controls, and the specification of the empirical model.

The majority of quantitative crisis- or dispute-escalation models take as their unit of analysis each of the dyads that make up the observed dispute, whether by selecting on disputing dyads from the outset or modeling the selection of pairs of states into disputes (e.g., Bennett and Stam 2004, Reed 2000). Call this a *participant-dyad* design, in which any given crisis or dispute may have several dyadic observations associated with it. For example, in conventional empirical models, ICB Crisis #342 pitting Libya against both France and Chad might consist of three dyads: Libya-France, Libya-Chad, and Chad-France. If the crisis were to escalate to war, the first two dyads would be coded in the affirmative, while the latter would not. Looking at the crisis as a collection of more or less independent dyads, this coding scheme would surely be descriptively accurate in terms of the participants, but apart from typical methods of clustering standard errors on individual crises, the conventional approach, whether it includes all dyads or drops "joiners," does not take into account the strategic dependence across these observations implied by this chapter's theory of coalitional crisis bargaining.

A simple example suffices to make the point. Consider the picture of the crisis-level distribution of power painted by the participant-dyad approach.[17] First, while Libya's military capabilities would be accurately represented, neither France's nor Chad's would be, as individual measures would not take

---

[17] See also Croco and Teo (2005), as well as Powell (1999, pp. 109–110).

into account the expected combination of their capabilities against Libya. Second, the crisis-level distribution of power – say, Libya's relative to its opponents – would also be incorrectly measured. A test that disaggregated this crisis into state-dyads would see one highly imbalanced pair (France-Libya) and a relatively balanced pair (Chad-Libya) of states nonetheless constrained to take on the same value of the outcome variable. At best, this introduces noise into the analysis and, at worst, could lead to suspect inferences over the effects of relative military power due to what is effectively an invalid measure of the concept. Thus, coalition-level variables such as aggregate military power, relative military power, preference diversity, and the like cannot be accurately measured and integrated into the standard participant-dyad research design.

To remedy this and to create a more appropriate research design on which to test the predictions of the theory, I construct a directed *side-dyad* dataset, which keeps the crisis as the unit of observation but simply aggregates relevant variables for the members of each side, whether singletons or coalitions. Specifically, I break each crisis into two observations, where each side has a chance to be observed as a coalitional side involving L (side 1) and as the target side involving T (side 2). Thus, Crisis #342 would produce two directed observations, one in which Chad (as L) and France (as P) are side 1 versus Libya's side 2, and another in which Libya is side 1 to Chad and France's side 2. This allows for an appropriate measure of both each side's military power and the presence of coalitions on either side of the crisis, which the standard participant-dyad approach is not designed to do. This results in a sample of 522 directed side-dyads, although missing data reduces the samples used for estimating the empirical models to 519.

The outcome variable is War, which equals one if the crisis in question escalated to "full-scale war" according to the ICB data, which involves large-scale and sustained violent clashes, and zero otherwise (Wilkenfeld and Brecher 2010). Of the 261 crises in the sample, 35 escalate to war, while the remainder are either resolved peacefully or escalate to a lower level of violence. Next, the theory implies two main theoretical variables: (a) whether side 1 is a coalition or a singleton and (b) the military capabilities of the target. Thus, I include $Coalition_1$, an indicator that equals one if side 1 is a coalition as defined in Chapter 2, and $CINC_T$, which again uses the Correlates of War Project's CINC data as a proxy for military power (Singer, Bremer, and Stuckey 1972, Singer 1987). Since coalitions often form – and negotiate their demands – in the shadow of some uncertainty about the presence of a coalition on the other side, I do not measure the total capabilities of *all* states on the opposing side, focusing instead on T, the primary belligerent. I do, however, include the presence of a coalition on side 2 as a control. Finally, since the theory makes predictions about both coalitions and singletons on side 1, I include the interaction term $(Coalition_1 \times CINC_T)$, facilitating a comparison of the effects of the target's military strength across both coalitional and singleton cases.

To rule out potential confounding factors, the model also includes several control variables. First, side 1's own aggregate military capabilities ($CINC_1$) and the number of states on side 1 ($Number_1$) are both correlated with the presence of a coalition and may also influence the escalation of crises to war. Further, by accounting for these factors, the remaining effects of the coalition variable are more likely to be driven by preference diversity, not simply numerical or military size. Other controls, all potentially linked to both coalition-building and war, are the COW level of contiguity between L and T (Stinnett et al. 2002), the percentage of the states on side 1 that share either defense or consultation pacts according to the Alliance Treaty and Obligations Project data (Leeds et al. 2002), two variables that indicate whether L and T are democracies according to the Polity IV scale (Marshall and Jaggers 2009), as well as in indicator that side 1 received the explicit support of the UN Security Council for side 1 (see Ch. 3 and Chapman 2011).[18] I also include two more dummy variables, one indicating the presence of a coalition on side 2 and another, as in Chapter 3, dividing the years after the end of the Cold War to account for potential differences in the ease of cooperation across periods of bi- and unipolarity (Voeten 2005, Waltz 1979).

The full statistical specification, then, is

$$\Pr(\text{War} = 1) = \Phi(\alpha + \beta_1 \text{Coalition}_1 + \beta_2 \text{CINC}_T$$
$$+ \beta_3 \left(\text{Coalition}_1 \times \text{CINC}_T \right) + \beta \mathbf{X}_{si} + \varepsilon_i), \tag{4.7}$$

where Φ is the CDF of the standard normal distribution, implying a probit model, and $\mathbf{X}_{si}$ is a vector of control variables measured by side s and crisis i. I also estimate Huber-White robust standard errors to account for intergroup correlations and cluster them by the individual crisis to account for the double-counting of each directed side of the crisis.[19] As noted earlier, the empirical predictions are derived from a strategic model where selection into coalitions is endogenous, which helps obviate concerns about selection bias, in addition to the selection model in the appendix to Chapter 3, which found no significant level of correlation in the errors between crisis onset and coalition formation.

4.3.3 Results: Crisis Escalation

Table 4.1 presents the results of estimating two empirical models. Model 1 uses the two theoretical variables and full complement of controls, while Model 2

---

[18] For the distance measure, I follow COW categories for decreasing levels of contiguity, 1–5, then code all states separated by more than 400 miles (category 5) as 6; results are similar if I simply use an indicator for land contiguity. Similar models using data from the Correlates of War alliance data (Gibler and Sarkees 2004) produces similar results to the ATOP-based measure.

[19] Results are substantively similar when standard errors are clustered by the lead state, as well as when the model is estimated with random effects that calculate a specific intercept for each year in which a crisis occurs.

TABLE 4.1. *Probit Models of Crisis Escalation, 1946–2001*

	Pr(War = 1)	
Variable	Model 1 No Interaction	Model 2 With Interaction
– *Theoretical variables* –		
$Coalition_1$	1.24 (0.32)***	1.31 (0.32)***
$CINC_T$	−5.60 (2.04)***	−4.32 (2.22)*
$Coalition_1 \times CINC_T$	--	−8.74 (4.91)*
– *Control variables* –		
$CINC_1$	−3.14 (1.28)**	−2.47 (1.44)*
$Number_1$	0.16 (0.09)*	0.14 (0.09)
$Percent\ Allied_1$	−0.87 (0.55)	−0.80 (0.57)
$Democracy_L$	0.09 (0.16)	0.11 (0.17)
$UNSC\ Support_1$	0.52 (0.32)	0.49 (0.32)
$Distance_{L,T}$	−0.07 (0.05)	−0.07 (0.05)
$Democracy_T$	0.33 (0.16)**	0.33 (0.17)**
$Coalition_2$	1.11 (0.21)***	1.12 (0.21)***
Post-Cold War	−0.22 (0.28)	−0.23 (0.28)
Intercept	−1.50 (0.23)***	−1.52 (0.23)***
	Model Statistics	
N	519	519
Log-likelihood	−166.28	−164.28
$\chi^2_{(d.f.)}$	$45.41_{(11)}$***	$56.53_{(12)}$***

Significance levels : ∗ : 10% ∗∗ : 5% ∗ ∗ ∗ : 1%

introduces the interaction term, ($Coalition_1 \times CINC_T$). Beginning with Model 1, crises with coalitions on side 1 are significantly more likely to escalate to war than those with singletons, just as they are in the bivariate relationship captured in Figure 2.3, and the relationship remains statistically discernible at $p < 0.01$, even with a full complement of control variables. Likewise, the target's military capabilities also appear to have a negative and statistically discernible effect on the probability that the crisis escalates to war. However, Model 1 cannot reveal how much of that negative relationship can be attributed to singletons or coalitions on side 1 – and, as a result, how much of the relationship between target strength and war is overstated by failing to account for the presence of coalitions in empirical analyses of crisis and dispute escalation.

Moving to Model 2, which introduces the interaction between the presence of a coalition and the target's military strength, the coalition variable grows slightly in magnitude and retains its statistical significance ($p < 0.01$). However, as the component of an interaction term, this coefficient can only

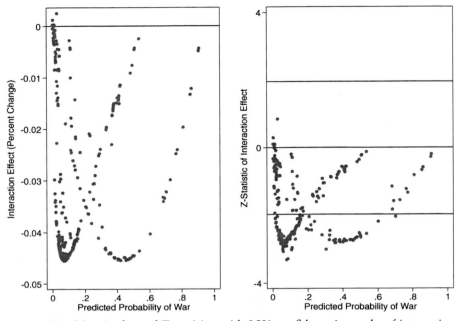

FIGURE 4.4. Magnitudes and Z-statisics, with 95% confidence intervals, of interaction effect by predicted probability of war

be interpreted as the effect of coalition-building when $CINC_T = 0$, or when the target is powerless. The coefficient on $CINC_T$, however, is meaningful for $Coalition_1 = 0$, capturing a case in which side 1 is a singleton. Hypothesis 4.1 predicts that this coefficient should be negative, if small, although the consistency of the sign should be weighed against the weaker statistical discernibility ($p < 0.1$) that comes along with accounting for the presence of a coalition on side 1. Next, the coefficient on the interaction term ($Coalition_1 \times CINC_T$) is both negative and statistically significant at $p < 0.1$, as predicted by Hypothesis 4.2. In nonlinear models like the probit employed here, however, neither the magnitude nor the statistical significance of the interaction effect can be assessed by the coefficient and standard error estimated for the interaction term itself (Ai and Norton 2003, Karaca-Mandic, Norton, and Dowd 2012).

Since Hypothesis 4.2 predicts different marginal effects of target capabilities for singletons and coalitions, I take two steps to assess the magnitude and statistical significance of the interaction effect. First, I use Ai and Norton's (2003) procedure to estimate the interaction effect for every observation in the data, then plot those effects and their test statistics against the predicted probaility of war estimated for each observation in Figure 4.4. Beginning in the left panel, the interaction effect is the difference of the marginal effect of $CINC_T$ on the probability of war between singletons and coalitions. Since CINC scores

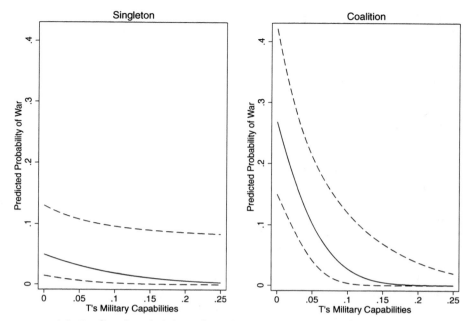

FIGURE 4.5. Predicted probability of war by presence of coalition and target military capabilities with 95% confidence intervals

are a state's percent share of all capabilities in the international system (Singer, Bremer, and Stuckey 1972), the natural unit of increase is a single percent of global capabilities, and the left panel has been scaled to represent the percent change in the probability of war associated with just such a one-unit increase in $CINC_T$. On average, targets that control 1 percent more of global capabilities will be less war-prone by roughly 0.04 percent.[20] More broadly, the interaction effect is negative for a sizable majority of observations, consistent across both low and high predicted probabilities of war. Unsurprisingly, however, more observations fall near the low end of the predicted probability of war, since peaceful outcomes are far more common than war in the sample. Moving to the right panel, which plots a z-statistic against 95% confidence intervals, we can see that the interaction effect is statistically discernible from zero for the majority of observations as well. Therefore, as predicted in Hypothesis 4.2, the marginal effect of increasing target capabilities is larger for coalitions than for single disputants, and the effect is generally statistically significant at $p < 0.05$.

Second, to gain some more substantive purchase on the interaction effect, Figure 4.5 presents the results of two simulations, plotting the effect of the target's military capabilities on the predicted probability of war for the "average"

---

[20] I calculated the average interaction effect by using Norton, Wang, and Ai's (2004) inteff package in Stata 13.1.

singleton and the "average" coalition on side 1, with 95% confidence intervals.[21] At the left of each horizontal axis, the target is virtually powerless as its CINC score approaches zero, but as it approaches 0.25, the target is equivalent to a superpower. In the left panel, where side 1 is made up only of a single state, the mean predicted probability of war begins around 0.05 and declines as the target grows stronger, and as predicted by Hypothesis 4.2, it falls quite slowly, and while the confidence interval generally declines, it is mostly flat, suggesting a weak relationship between target capabilities and the probability of war for singletons. However, moving to the right panel, where side 1 is a coalition, there is a substantively large and statistically significant negative relationship between the target's military strength and the probability of war. For the weakest targets, the mean predicted probability of war is roughly 0.275, and this probability falls rapidly as the target grows more powerful, falling below the singleton's predicted probability of war at roughly $CINC_T = 0.12$.

In addition to being consistent with Hypotheses 4.1 and 4.2, Figure 4.5 can also help rule an additional alternative interpretation of the theoretical variables. Arguments about coalitional signaling ineffectiveness (Christensen 2011, Lake 2010/2011), as well as some conjectures about alliances and spirals of hostility (Vasquez 2009), imply that coalitions increase the probability of war unconditionally: whether by virtue of the diversity of preferences inside them, the prospect of collective failures to apply military pressure, or a link between military cooperation and the "security dilemma" (Herz 1951, Wagner 2007), coalitions should fail to convey resolve effectively, however weak or strong their targets might be, leading to a greater probability of war than singletons. If this were the case, then the predicted probability of war in the right panel should be both flat and everywhere higher than the predicted probability in the left panel, but this is clearly not the case. Rather, as predicted by the theoretical model, the coalitional probability of war can be higher or lower than the singleton probability of war, and its relative size is a function of the target's military strength.

In practice, then, coalitions do not appear universally inept at signaling their intentions, nor are they unconditionally more likely to end up in wars than states acting alone, as suggested by the simple bivariate relationship found in Chapter 2. To the contrary, the theoretical model suggests that coalition partners might facilitate effective signaling by discouraging bluffing against strong targets. This offers a plausible account – discussed in more detail in the following section – of both moderated threats and the peaceful resolution of the Berlin Crisis of 1961–1962, as the United States was discouraged from embarking on risky demonstrations of resolve (involving military deployments across the intra-German border) by allies that could credibly promise not to participate in an ensuing war. On the other hand, as evidenced by the higher

---

[21] All continuous variables are held at their means, while all dummy variables are held at their modes.

probability of war against weak targets, coalitions may send weak signals when lead states moderate their threats to preserve military cooperation – raising the probability of war above what it would be if they acted unilaterally or refused to accommodate the demands of their coalition partners. This may explain both Serbia's refusal to acquiesce in the face of NATO threats before the Kosovo War as well as the United States' willingness to mask its resolve and run the risk that Serbia would find its threats incredible.

4.4 APPLICATION: THE BERLIN AND KOSOVO CRISES

Why are some coalitions' military threats met with resistance from their targets, while others lead to peaceful resolutions of the underlying issues? The crises over Berlin and Kosovo, in 1961–1962 and 1998–1999, respectively, are particularly useful for answering this question for several reasons. First, these crises are similar in key respects; both involve powerful coalitions led by the United States, composed of major but skittish military powers in NATO; further, each coalition's primary target was uncertain over the United States' resolve. Second, they differ in both outcome – peaceful resolution over Berlin, an extended air campaign over Kosovo – and one of the theory's key explanatory variables: the military strength of the target. Since factors constant across the cases cannot be used to explain different outcomes, this helps us focus on the why and the how of the relationship between the target's strength, signaling dynamics, and the escalation of crises to war. As such, I show in this section how the incentives to preserve military cooperation in the face of partners' skittishness can produce credible signals by discouraging bluffing in some crises, such as Berlin, and incredible signals that fail to convey resolve and lead to war in others, as happened over Kosovo.

Before discussing the cases, it is important to note that some actors in the theory, the historical figures in the cases, *and* the analyst are all uncertain over a particular quantity: in this case, a lead state's resolve, or its willingness to use force in lieu of accepting a given settlement. As a result, establishing precisely what settlements the United States would prefer to war over either Berlin or Kosovo is difficult, given actors' basic incentives to misrepresent their resolve, whether to their adversaries (Fearon 1995) or to their interviewers or biographers ex post.[22] However, by formalizing the theory, I can derive predictions over the effects of factors that are more easily observed, for example military capabilities and partner skittishness, that do not depend on particular levels of uncertainty. In other words, the theory's predictions take the unobservability of preferences and incentives to misrepresent them explicitly into account,

---

[22] This is most certainly true for biographies and memoirs, which are often themselves – especially in edited and published form – carefully calculated pieces of political speech. Clark's (2012) discussion of the myriad national histories of the origins of the First World War is especially telling on this point.

relating them to observable factors like military capabilities. Thanks to the precise connections drawn between premises and conclusions in the theory, we can analyze the cases of Berlin and Kosovo despite uncertainty over the true levels of American resolve.[23]

It is possible, however, to make some limited judgments ex post about actors' resolve by updating our beliefs according to Bayes' Rule, just as the target state does.[24] For example, if war breaks out, the model tells us that it is because L proves more resolute than T expects; otherwise, if the lead state were irresolute, it would have accepted the same offer in lieu of war. Peace, however, can prevail for any number of reasons, from successful signaling of L's true level of resolve to the target's beliefs over L's willingness to fight. Since settlements can be reached by both resolute and irresolute types, it is difficult to determine the content of L's private information unless we know more about (a) the equilibrium being played, whether separating, pooling, or semi-separating; (b) the proposals that T makes, relative to those it considered but opted against; and (c) the size of the coalition's military threat, likewise relative to those it considered but opted against. Therefore, we can confidently say below that the United States and its partners *were* resolute over Kosovo, but its true level of resolve over Berlin cannot be judged simply from the fact that war was averted. To do so would commit the fallacy of affirming the consequent: (a) if the United States were irresolute, the Berlin Crisis would be resolved peacefully; (b) the Berlin Crisis was resolved peacefully; (c) therefore, the United States was irresolute. This is clearly incorrect reasoning, since the conclusion does not follow from the premises: the crisis might also have been resolved peacefully if the United States were resolute. A more complete model, such as the one presented here, helps us anticipate and avoid this fallacy. Inferring true levels of resolve, however, is less important than telling a plausible story about the effect of target strength on the credibility and effectiveness of signals and military threats, to which I now turn.

The Second World War in Europe ended with the division of Germany into zones of Allied occupation, which had hardened by 1961 into two new states: the Federal Republic of Germany (FRG, or West Germany), occupied by the United States, United Kingdom, and France; and the German Democratic Republic (GDR, or East Germany), occupied by the Soviet Union.[25] The former capital, Berlin, was similarly divided, though the democratic Western half was

---

[23] All theories of politics rest on unobservable quantities. However, some are better designed to make predictions in the face of such unobservability than others.

[24] For a similar exercise applied to trust and reciprocation games in the Cold War, see Kydd (2005).

[25] France's inclusion may at first seem puzzling, given its rapid conquest in 1940 (May 2000), but even as soon as the Yalta conference, the United Kingdom and United States were eager to restore France as a great power in anticipation of postwar fractures in the Allied coalition (Plokhy 2010, Riker 1962). For more on the dissolution of war-winning coalitions, see Chapter 5.

isolated in the authoritarian East. While the four-power agreements among the Allied victors guaranteed free movement throughout the city, Berlin come to be viewed by East German leader Walter Ulbricht as a serious threat to the GDR (Kempe 2011, pp. 92–94), as thousands of his citizens had already escaped to the relative liberty of the West, draining his country of badly needed human capital. With the viability of the communist experiment in Germany – and the Soviet Union's reputation in the global communist movement generally – at stake, Soviet premier Nikita Khrushchev had resolved by the early months of 1961 that the time to alter the status quo in Berlin had come (Fursenko and Naftali 2006, p. 353).[26]

The crisis began during the Soviet-American Vienna Summit on 4 June 1961, when Khrushchev announced to the new American president John F. Kennedy that he would sign a unilateral peace treaty with East Germany and commit the Soviet Union to its defense. Such a move would invalidate the four-power agreements that were made as part of Nazi Germany's 1945 surrender and allow East Germany the right to close its border with and control access to the West – particularly in Berlin – as it saw fit. Faced with an American president mere months in office and scrambling to recover some prestige after the Bay of Pigs fiasco (Dallek 2003, Kempe 2011), Khrushchev had ample reason to suspect that Kennedy was less resolute than his predecessor, former Supreme Allied Commander Dwight Eisenhower (Fursenko and Naftali 2006, ch. 14). This uncertainty incentivized him to probe the young president's resolve over other issues, like the status of Berlin (see Wolford 2007). Thus, while American resolve over West Berlin was uncertain, Khrushchev had reason to be optimistic that the new administration's stance would be more flexible than Eisenhower's, who demonstrated his own commitment to guaranteeing the status of West Berlin in a similar crisis three years earlier.

The American response emerged only after considerable deliberation both inside the Kennedy administration and with key NATO allies – members of any American-led coalition that might find itself engaged against the Warsaw Pact over the fate of Germany. While France and West Germany opposed any signs of conciliation, lest the Soviets be encouraged to act unilaterally against Berlin, discussions between American and British diplomats ultimately settled on a "double-barreled" (Aono 2010, pp. 326, 334) strategy combining a limited, delayed military mobilization and, for the first time, a rhetorical distinction between "East Berlin" and "West Berlin" (Kempe 2011, pp. 312–314).[27] While French and West German concerns over the weakness of the threat were largely dismissed as posturing – and no threat to their coalitional commitment in any

---

[26] By the early 1960s, Khrushchev's strategy of "peaceful coexistence" faced a vigorous challenge from within the Communist camp in the form of strident Chinese criticism, and the approach of a major Party Congress in the fall gave Khrushchev added incentive to attain a favorable resolution of the Berlin question during the summer (see Kempe 2011).

[27] See also Freedman (2000, p. 71).

case – British skittishness did prove to be a factor in choosing a signal that, in essence, communicated that the United States was unwilling to fight over Berlin *if* the Soviets and the GDR merely did what they wished in their own zone of occupation (Freedman 2000, p. 67).

The crisis headed toward resolution when GDR military forces began construction of a barrier separating East and West that would eventually become the Berlin Wall, stanching the flow of refugees that had been steadily pouring out into the West. Tensions would continue for months, as Soviet troops wargamed an assault on West Berlin and Khrushchev unilaterally ended a moratorium on nuclear testing (Kempe 2011, p. 453), but a settlement would eventually stabilize around a limited revision to the status quo, since the four-power agreements between the occupying powers explicitly forbade GDR troops from operating inside Berlin. Thus, the great powers stood at the brink of a third world war and found a solution that preserved the peace, in large part because Khrushchev ultimately believed that the relatively small fait accompli of sealing off West Berlin was insufficient to drive the United States to war (Kempe 2011, p. 382). Given the Soviet Union's military strength, the Americans would be highly unlikely to alienate coalition partners with a strategy of risky bluffing. In the language of the theoretical model, it appears that the United States chose a small level of mobilization, honestly revealing a lack of resolve over minor changes to the situation in Berlin, following the logic of a separating equilibrium when faced with a strong target. Since the cost of losing coalition partners was greater than the costs of bluffing, the need to secure a coalition partner's cooperation facilitated a peaceful settlement of what would be the most serious Cold War crisis until Soviet missile installations were discovered in Cuba in 1962.

If the model is to provide a credible explanation of the case, however, it should be consistent with several more facts. Specifically, the United States must have chosen a small mobilization in the interests of avoiding bluffing and, critically, retaining the support of coalition partners. In fact, Dean Acheson, recruited as an advisor to help a Kennedy administration widely believed to be struggling in foreign policy, advocated sending a battalion-level contingent of ground forces across the intra-German border and up the Autobahn as a show of resolve (Freedman 2000, pp. 62–63, 66–67), an "ominous thing" for the Soviets wondering just how far the United States could be pushed (Kempe 2011, p. 336–337, 519). Doves in the administration, however, emphasized the need to escalate the crisis only to a level that would ensure allied cooperation in the event of war (Kempe 2011, pp. 277, 300, 475) – a clear nod to the concerns of the British, who Kennedy knew "had no stomach for troops storming up the Autobahn" (p. 282). France, though outwardly hawkish, was also pointed out to be "busy" in Algeria at the time (*ibid.*, p. 282). Ultimately, a United States unwilling to go to war unless the whole fate of Europe was in play, a fact that both Kennedy and Secretary of State Dean Rusk emphasized privately and to subordinates throughout the crisis (Kempe 2011, pp. 319,

528), bowed to the requests of its coalition partners to limit the size of its mobilization (Aono 2010).[28] Kennedy knew that a larger mobilization – for example, Acheson's proposal – would have been a bluff (Kempe 2011, p. 558), and he valued maintaining British cooperation more highly than the potential gains of convincing the Soviets of a higher level of resolve in what appears to be a clear analog to the separating equilibrium of the skittish partner case identified in the theoretical model.

While the Berlin Crisis ended peacefully, another American-led coalition in 1998-1999 issued a military threat against Serbia, lest it withdraw army units from the breakaway region of Kosovo, only to see the target state refuse to comply and call in the threat of war. Operation Allied Force, which began on 24 March, 1999 after coercive diplomacy failed to win Serbian acquiescence, would see a massive air war launched by NATO in its first-ever collective military operation. The seventy-eight-day war would eventually see Serbian troops withdrawn from Kosovo, where conflict with the local Albanian minority was verging on ethnic cleansing before NATO chose to launch the war. NATO had threatened Serbian president Slobodan Milosevič with an air war unless he withdrew Serbian army units from Kosovo as early as October 1998 (Henriksen 2007, p. 149), the orders for which had been standing for months as the alliance announced publicly its mobilization process, drawing contributions from across the alliance for an air campaign that would be launched chiefly from bases in Italy and carriers at sea. The beginning of the war followed months of fruitless negotiations between Serbian leaders and NATO officials, yet the threat of military punishment was insufficient to compel Milosevič's compliance; for that, NATO had to go to war.

The air campaign would eventually move from Kosovo itself to political and military targets in the Serbian capital of Belgrade, destroying Serb military capabilities and ultimately setting the stage for the social unrest that would sweep Milosevič from power in the years following the war. Clearly, acquiescing to NATO's initial threats would have left Milosevič in a better position: with his troops out of Kosovo but untouched by foreign air power, his political power and personal safety might have remained intact. However, it is clear that he did not believe NATO's threat to wage a long, costly air campaign, much less one that would eventually see NATO making moves toward a ground invasion through Albania (Clark 2001, Daalder and O'Hanlon 2000). In the fall of 1998, in fact, Milosevič reportedly said to a group of aides, "So what? First they bomb, then peace resumes" (Henriksen 2007, p. 150). Ex post, his beliefs about NATO's resolve proved inaccurate, and the relevant question is *why* NATO threats were ineffective in changing the Serbian leader's beliefs about the coalition's willingness to wage a protracted campaign – that is, why NATO's threats were not credible.

---

[28] See also Kempe (2011, p. 518), for evidence that the British viewed their entreaties for limited mobilization as just this kind of restraint.

War occurs in the theoretical model when a resolute lead state fails to convince its target that it is indeed willing to fight rather than accept a wide range of potential bargains. This was certainly the case in the Kosovo war, but the reason that NATO was unable to signal its resolve is *not* the problem posed by irresolute types and their incentives to bluff. Separating oneself requires a costlier signal, and NATO clearly had such signals available – the British, for example, consistently advocated mobilizing for a ground war as well as an air war (Richardson 2000) – yet chose not to send them. Rather, NATO masked its resolve by making a threat consistent with what an irresolute type, one only willing to fight a short war, would make. Thus, it looks as though NATO and Serbia found themselves in the pooling equilibrium of the three-player game, such that a resolute type chooses a small mobilization, and its target has no reason to alter its beliefs as a result. Specifically, the target is aware that a resolute lead state would moderate its threats to keep the coalition together, and it retains its prior beliefs.

The details of the crisis, both inside NATO and in Serbia, are consistent with these predictions from the model. For example, even after the killing of forty-five Kosovar Albanians at Račak, several coalition members remained unwilling even to consider an air campaign, prompting significant political pressure from the United States and the United Kingdom (Daalder and O'Hanlon 2000, pp. 72, 73). Italy, from which numerous NATO aircraft would launch on their missions over Serbia, was particularly hesitant thanks to popular anti-Americanism and an awareness that proximity to the conflict zone would render it uniquely vulnerable to spillovers or retaliation (Cremasco 2000, p. 170). Likewise, German public opinion was supportive of threatening an air war but firmly opposed to a ground war (Rudolf 2000, pp. 134–138), and Spain was particularly dovish early on, given its own problems with separatist groups at home (Haglund 2000). As a result, "the United States sought to refine its proposed strategy in a way that would garner NATO support," including the identification of clear political actions that could end the fighting, limiting the threat to an air war, and a commitment (albeit private) to participate in a subsequent stabilization force, which would minimize the costs of participation for nearby skeptics like Italy and Greece (Daalder and O'Hanlon 2000, p. 73).

Tellingly, more hawkish members of the coalition were aware of the problems of choosing a small mobilization. Frustrated by NATO's limited military threat, British Prime Minister Tony Blair lamented the effect on Milosević's estimate of the coalition's willingness to fight: "He thinks we're unwilling to act. If he thought there was real steel in the threat he wouldn't get away with it" (Richardson 2000, p. 146). Aware of the balancing act required to harmonize NATO's aims and that the difference between high and low mobilizations against his own relatively weak country was small, Milosević reasonably inferred that the United States would preserve the coalition rather than lose it in return for a higher mobilization that was itself not much more costly than war with its partners. Thus, he retained prior beliefs that indicated a serious lack of coalitional resolve.

He had good reason to believe that the air campaign would not be sustained, considering the U.S. missile operations against Osama Bin Laden during summer 1998, U.S. and British air strikes against Iraq in December 1998, and the well-known reluctance by several NATO countries even to begin limited air strikes in the period leading up to [Operation Allied Force]. (Henriksen 2007, 157)

With no reason to change his beliefs in the pooling equilibrium, Milosević resisted the coalition's demands, and the result was nearly three months of escalating air attacks, ending only when the threat of ground action and Russian diplomatic pressure finally induced Serbia to yield.[29]

As these crises demonstrate, skittish partners can sometimes be a hindrance to effective, credible signaling, particularly against weaker targets where the mobilization gains of sending stronger threats are smaller than the value of retaining coalition partners. Just such a problem led NATO to send a "weak" signal to Serbia in 1999. However, against a more formidable opponent in 1961, the United States kept its coalition together, honoring the wishes of a skittish partner by choosing a small mobilization, signaling credibly that it was unwilling to go to war over small changes to Allied rights in Berlin; as a result, the Berlin Crisis ended peacefully and with the Western Alliance intact. Thus, skittish coalition partners can facilitate war or peaceful settlement, and whether they do so depends on whether moderation is the result of a discouraged bluff or a trade-off between demonstrating resolve and keeping the coalition intact.

4.5 SUMMARY AND DISCUSSION

Whether and when crises escalate to war is one of the most difficult, and thus most enduring, questions in the study of international relations. Yet despite the prevalence of coalitions in some of the more consequential conflicts of the last century, their role in crisis escalation has remained poorly understood. The theoretical model in this chapter focused on the second stage of military multilateralism – bargaining with a target state under asymmetric information – to illuminate the strategic linkages between intra-coalitional politics on the one hand and threats and signaling on the other. When lead states must compensate skittish partners for valuable military cooperation, partners play a surprising role in the course and resolution of crises. If a skittish partner can threaten to deny military cooperation over the lead state's mobilization decisions, the lead state must choose between signaling resolve and ensuring military cooperation, and how it solves this trade-off depends on the military strength of

---

[29] See also Daalder and O'Hanlon (2000, pp. 94–95), who also emphasize that NATO's failure to follow through on previous threats also encouraged Milosević to doubt its members' resolve.

the target. Therefore, potential coalition partners introduce commitment problems that bear on whether a dispute's primary belligerents can solve their own information problems.

These dynamics emerge from the effect of the partner's commitment problem on the costs of bluffing. First, when the target is not too weak, an irresolute lead state is not tempted to bluff, preferring to maintain military cooperation with the partner, building a coalition and extracting what concessions it can from the target peacefully. In this way, the partner's presence leads to a weaker or less serious threat, an oft-noted feature of coalitions' attempts at coercion (Byman and Waxman 2002), but the alternative to a modest threat is a war-risking attempt to bluff about one's resolve. In other words, ostensibly "weak" threats help preserve the peace, ensuring an efficient outcome, where states would otherwise be willing to risk war. At the same time, truly resolute states may find unilateral action an effective way to signal their own resolve against powerful targets, facilitating peaceful settlement by virtue of foregoing coalitional action. Second, when the target is sufficiently weak, otherwise resolute lead states may be tempted to send weaker signals, limiting mobilization in order to preserve military cooperation at the cost of fighting a war that could have been avoided with higher levels of mobilization. This is consistent with the popular image of coalitions as producing "watered-down" threats and sending weak signals of resolve. Previous scholarship has speculated that interested outsiders can interfere with otherwise straightforward incentives to signal resolve (Fearon 1997, Russett 1963). The preceding analysis shows that these conjectures can be reproduced in a deductive model, and it makes predictions over when this elevated probability of war is likely to occur: when coalition members differ in their sensitivity to the costs of war and the coalition's target is not too strong. Further, despite the optimality of cooperation, it nonetheless comes with a price tag of raising the probability of war above what it would be if there were no opportunities to build a coalition. Put more starkly, the promise of cooperation can increase the probability of war, because it would not occur if coalition partners were not available.

The empirical analysis of crisis escalation from 1946–2001 takes advantage of a research design, the side-dyad approach, that is uniquely suited to accurate measures of crisis-level variables. The results are consistent with the theory's predictions, in that (a) the probability of war does not depend on the target's strength when states act unilaterally, but (b) the probability of war is greatest against weaker targets when states build coalitions. Thus, contrary to a number of common conjectures about coalitional politics (Christensen 2011, Lake 2010/2011), coalitions may be more *or* less prone to war than states acting alone, and whether they are depends on both the distribution of preferences within the coalitions and strength of their targets, critical factors omitted from most extant work on the topic of coalitions and war. To make the point again, coalitions are not merely the sum of their parts; from the perspective of signal-solve, they are very often less but sometimes quite a bit more. Further, the logic of the model appears evident in two prominent coalitional crises, over Berlin during the Cold War and over Kosovo in the last years of the twentieth century.

Finally, the chapter builds on the general literature on signaling, bargaining, and crisis escalation by providing a substantive example for a crucial and often overlooked second element of Fearon's (1995) rationalist explanation for war based on private information: incentives to misrepresent one's resolve or willingness to fight. Absent a reason to dissemble, bargains are easy to reach as long as players are patient (Leventoglu and Tarar 2008), and explaining observed conflicts in terms of this mechanism requires an account of why belligerents were unwilling or unable to share war-avoiding information. Other incentives to misrepresent, be they secrecy about military strength (Meirowitz and Sartori 2008) or bolstering a reputation for resolve (Wolford 2007), manifest as irresolute types mimicking resolute types. However, the story presented here involves a resolute state choosing not to reveal itself as such, even when the dyadic characteristics of the crisis are conducive to credible signaling. This suggests both coalition-specific avenues for understanding the causes of war as well as challenging policy questions over the trade-off between building foreign policy coalitions and aiming effective compellent threats at one's adversaries.

4.6 APPENDIX

Proof of Proposition 4.1. I first prove the existence of separating and semi-separating equilibria, then show that pure pooling equilibria, on both high and low mobilizations, do not exist.

The separating equilibrium occurs when $m_T \geq \hat{m}_T$. Strategies and beliefs are as follows. $\overline{v}_L$: choose $m_L^* = \overline{m}_L$ and accept iff $x \geq \overline{x}$. $\underline{v}_L$: choose $m_L^* = \underline{m}_L$ and accept iff $x \geq \underline{x}$. T: believe $\phi' = 1$ if $m_L^* = \overline{m}_L$ and propose $x^* = \overline{x}$; believe $\phi' = 0$ if $m_L^* = \underline{m}_L$ and propose $x^* = \underline{x}$. Since each possible mobilization level is chosen, there are no out-of-equilibrium beliefs.

To prove the existence of this equilibrium, we first derive each type of L's optimal mobilization level given its type. The resolute L chooses $\overline{m}_L$ such that it satisfies

$$\max_{m_L}\{-m_L + \frac{m_L}{m_L + m_T}\overline{v}_L - c_L\}.$$

The optimum satisfies the first-order condition, $m_T\overline{v}_L/(m_T + m_L)^2 - 1 = 0$, yielding

$$\overline{m}_L = \sqrt{m_T}\sqrt{\overline{v}_L} - m_T. \tag{4.8}$$

Substituting $\overline{m}_L$ into the second partial with respect to m_L yields $-2/\left(\sqrt{m_T}\sqrt{\overline{v}_L}\right) < 0$, ensuring that $\overline{m}_L$ is a maximum. Since player-types' payoffs differ only in their valuations v_L, we can also define the irresolute type's optimum as

$$\underline{m}_L = \sqrt{m_T}\sqrt{\underline{v}_L} - m_T,$$

where $\underline{m}_L < \overline{m}_L$, since $\underline{v}_L < \overline{v}_L$. Next, consider what proposals L accepts after choosing m_L. The resolute type accepts iff x satisfies

$$x\overline{v}_L \geq \frac{\overline{m}_L}{\overline{m}_L + m_T}\overline{v}_L - c_L \Leftrightarrow x \geq 1 - \frac{c_L + \sqrt{m_T}\sqrt{\overline{v}_L}}{\overline{v}_L} \equiv \overline{x}, \qquad (4.9)$$

and the irresolute type's range of acceptable proposals can be defined by the same process as

$$x \geq 1 - \frac{c_L + \sqrt{m_T}\sqrt{\underline{v}_L}}{\underline{v}_L} \equiv \underline{x}, \qquad (4.10)$$

where, again, $\underline{v}_L < \overline{v}_L$ ensures that $\underline{x} < \overline{x}$.

Next, we can verify that T proposes $x^* = \overline{x}$ if $m_L^* = \overline{m}_L$ and $x^* = \underline{x}$ if $m_L^* = \underline{m}_L$, given its beliefs defined above. First, when $m_L^* = \overline{m}_L$, T proposes $x^* = \overline{x}$ rather than some $x' \geq \overline{x}$, since it will win L's acceptance but leave T strictly worse off than $\overline{x}$. It also will not propose some $x' < \overline{x}$, provoking rejection, as long as

$$1 - \overline{x} \geq \frac{m_T}{\overline{m}_L + m_T} - c_T, \qquad (4.11)$$

which is guaranteed to be true since $c_L, c_T > 0$. Substituting $\underline{x}$ and $\underline{m}_L$ where appropriate also ensures that T will propose $x^* = \underline{x}$ following $m_L^* = \underline{m}_L$.

Finally, we verify that each player-type of L's prescribed action is incentive-compatible. First, the resolute type must choose $m_L^* = \overline{m}_L$ rather than deviate to $\underline{m}_L$, which convinces T that it is irresolute. If T believes L to be irresolute, then it offers terms that only the irresolute type will accept, forcing the irresolute type to reject in its return to optimal play. Therefore, the resolute L reveals its type via high mobilization when

$$-\overline{m}_L + \overline{v}_L\overline{x} \geq -\underline{m}_L + \frac{\underline{m}_L}{\underline{m}_L + m_T}\overline{v}_L - c_L \qquad (4.12)$$

which is true as long as war is costly, $\underline{v}_L < \overline{v}_L$, and $m_T < \underline{v}_L$ (where the latter ensures that mobilization levels are positive). Finally, the irresolute type must choose $m_L^* = \underline{m}_L$ rather than deviate to $\overline{m}_L$, which convinces T that it is resolute. Since it is sure to accept any proposal that the resolute type does, its payoff to mimicking the resolute type involves accepting a relatively favorable proposal, meaning that it chooses low mobilization when $-\underline{m}_L + \underline{v}_L\underline{x} \geq -\overline{m}_L + \underline{v}_L\overline{x}$, which is true when

$$m_T \geq \frac{c_L^2\left(\sqrt{\overline{v}_L} + \sqrt{\underline{v}_L}\right)^2}{\overline{v}_L\left(\sqrt{\overline{v}_L} - \sqrt{\underline{v}_L}\right)^2} \equiv \hat{m}_T \qquad (4.13)$$

Therefore, the proposed separating equilibrium exists when $m_T \geq \hat{m}_T$.

The semi-separating equilibrium occurs when $m_T < \hat{m}_T$. Strategies and beliefs are as follows. $\overline{v}_L$: choose $m_L^* = \overline{m}_L$ and accept iff $x \geq \overline{x}$. $\underline{v}_L$: with probability h, choose $m_L^* = \overline{m}_L$ and accept iff $x \geq \underline{x}_h$; $\underline{v}_L$: with probability $1 - h$, choose $m_L^* = \underline{m}_L$ and accept iff $x \geq \underline{x}$. T: believe $\phi' = \phi(h)$ if $m_L^* = \overline{m}_L$, then propose $x^* = \overline{x}$ with probability r and $x^* = \underline{x}_h$ with probability $1 - r$; believe $\phi' = 0$ if $m_L^* = \underline{m}_L$, and propose $x^* = \underline{x}$. Again, there are no out-of-equilibrium beliefs.

Mobilization levels $\underline{m}_L$ and $\overline{m}_L$ are as defined above, and the resolute type's disincentive to deviate to $\underline{m}_L$ remains the same, since the target would again believe L to be irresolute and make a proposal that the resolute type is sure to reject. Likewise, T's optimal response to observing $\underline{m}_L$ is identical to that derived above. What remains is to establish the range of proposals $x \geq \underline{x}_h$ that $\underline{v}_L$ accepts after bluffing, T's posterior belief $\phi' = \phi(h)$ in the event of observing high mobilization, $\underline{v}_L$'s probability of choosing $\overline{m}_L$ that renders T indifferent between risking war and buying off both types, and T's probability of risking war that renders $\underline{v}_L$ indifferent between bluffing and honest revelation.

First, after choosing high mobilization, the irresolute L has a new expected value for war, which determines the proposals it accepts in equilibrium. It accepts proposals that satisfy

$$x\underline{v}_L \geq \frac{\overline{m}_L}{\overline{m}_L + m_T}\underline{v}_L - c_L \Leftrightarrow x \geq 1 - \frac{\sqrt{m_T}}{\sqrt{\overline{v}_L}} - \frac{c_L}{\underline{v}_L} \equiv \underline{x}_h. \tag{4.14}$$

Since types differ only in their valuation for war, the irresolute L is sure to accept both this proposal and any more generous one, that is, $x \geq \overline{x}$, that the resolute type accepts. However, the resolute type will not accept $\underline{x}_h$, since its expected value for war is higher. This creates a risk-return trade-off for T if it observes high mobilization: it can propose $x^* = \underline{x}_h$, which $\underline{v}_L$ accepts and $\overline{v}_L$ rejects, or $x^* = \overline{x}$, which both accept. As in similar models, it is straightforward to show that, as long as one type accepts a proposal, the costliness of war ensures that T never makes a proposal that both reject.

Second, let $\underline{v}_L$ choose $m^* = \overline{m}_L$ with probability h. Then, by Bayes' Rule, T's posterior belief that L is resolute, given $m^* = \overline{m}_L$, is $\Pr(\overline{v}_L|\overline{m}_L) = \phi/(\phi + h(1 - \phi))$. Third, $\underline{v}_L$ chooses the probability of bluffing, h, to render the target indifferent over proposing $\overline{x}$ and $\underline{x}_h$, after observing high mobilization. Formally, h solves

$$\frac{\phi}{\phi + h(1 - \phi)}\left(\frac{m_T}{\overline{m}_L + m_T} - c_T\right) + \left(1 - \frac{\phi}{\phi + h(1 - \phi)}\right)(1 - \underline{x}_h) = 1 - \overline{x},$$

where $\Pr(\overline{v}_L|\overline{m}_L)$ is as defined above. This yields its equilibrium probability of choosing high mobilization, h^*, where

$$h^* = \frac{\phi(c_L + c_T\overline{v}_L)\underline{v}_L}{(1 - \phi)c_L(\overline{v}_L - \underline{v}_L)}.$$

Note that h^* takes on plausible values, $0 \leq h^* \leq 1$, when types are well defined, or when $\underline{v}_L \leq (c_L \bar{v}_L (1 - \phi)) / (c_L + \phi c_L \bar{v}_L)$. When types are well defined, T finds it profitable to potentially separate them with its proposal – a precondition for a genuine risk of war – so I maintain this assumption throughout the analysis.

Finally, rendered indifferent over its available proposals, T chooses a probability of risking war, r, that renders $\underline{v}_L$ indifferent over bluffing and honestly revealing its type. Thus, r must solve $-\underline{m}_L + \underline{x}(\underline{v}_L) = \overline{m}_L + r(\underline{x}_b, \underline{v}_L) + (1 - r)(\overline{x}\underline{v}_L)$, yielding the equilibrium probability with which T makes the risky proposal,

$$r^* = 1 + \frac{\sqrt{m_T}\sqrt{\bar{v}_L}}{c_L}\left(1 - \frac{2}{1 + \sqrt{\underline{v}_L}/\sqrt{\bar{v}_L}}\right).$$

Algebra shows that T's mixing probability takes on plausible values $0 < r^* \leq 1$ when $m_T < \hat{m}_T$, as defined above, which establishes $\hat{m}_T$ as the cutpoint dividing the equilibrium space between pure separating and semi-separating equilibria.

It remains to show that pooling equilibria in which both player-types choose the same action do not exist. First, the resolute L's refusal to pool with the irresolute type, established by Inequality (4.12), is strictly true given $\underline{v}_L < \bar{v}_L$ and any out-of-equilibrium beliefs T might hold, since L is sure to receive its reservation value after high mobilization regardless of T's strategy. Therefore, there can exist no equilibrium in which both types pool on low mobilization. Second, the irresolute type is unwilling to pool on high mobilization, since by assumption T's priors are sufficiently optimistic that it will risk war – that is, propose $\underline{v}_L$'s reservation value, which $\bar{v}_L$ goes on to reject – in a pooling equilibrium. For T to risk war, it must be the case that

$$\phi\left(\frac{m_T}{\overline{m}_L + m_T} - c_T\right) + (1 - \phi)(1 - \underline{x}_b) > 1 - \bar{x} \Leftrightarrow \phi < \frac{c_L(\bar{v}_L - \underline{v}_L)}{\bar{v}_L(c_L + c_T \underline{v}_L)} \equiv \phi_w.$$

To tolerate pooling under these conditions, the irresolute type would require $-\overline{m}_L + \underline{x}_b \underline{v}_L \geq -\underline{m}_L + \underline{x}(\underline{v}_L)$, which cannot be true as long as war is costly and $\underline{v}_L < \bar{v}_L$. Therefore, in the two-player case there are two equilibria: (1) a fully separating equilibrium when $m_T \geq \hat{m}_T$, and (2) a semi-separating equilibrium when $m_T < \hat{m}_T$. $\square$

Proof of Proposition 4.2. I first prove the existence of separating and semi-separating equilibria, then show that pure pooling equilibria, on both high and low mobilizations, do not exist.

The separating equilibrium occurs when $m_T \geq \hat{m}_T$. Strategies and beliefs are as follows. $\bar{v}_L$: choose $m_L^* = \overline{m}_L$ and accept iff $x \geq \bar{x}$. $\underline{v}_L$: choose $m_L^* = \underline{m}_L$ and accept iff $x \geq \underline{x}$. T: believe $\phi' = 1$ if $m_L^* = \overline{m}_L$ and propose $x^* = \bar{x}$; believe $\phi' = 0$ if $m_L^* = \underline{m}_L$ and propose $x^* = \underline{x}$. Since each mobilization level is chosen, there are no out-of-equilibrium beliefs. P: cooperate for all m_L.

To prove the existence of this equilibrium, begin by deriving each type of L's optimal mobilization level given its type and P's strategy. Since P cooperates for any mobilization choice, each type of L solves its maximization problem in the expectation of P's support. Thus, the resolute L chooses $\overline{m}_L$ such that it satisfies

$$\max_{m_L}\{-m_L + \frac{m_L + m_P}{m_L + m_P + m_T}\overline{v}_L - c_L\}.$$

The optimum satisfies the first-order condition, $m_T\overline{v}_L/(m_P + m_T + m_L)^2 - 1 = 0$, yielding

$$\overline{m}_L = \sqrt{m_T}\sqrt{\overline{v}_L} - m_T - m_P. \tag{4.15}$$

Substituting $\overline{m}_L$ into the second partial with respect to m_L yields $-2/\left(\sqrt{m_T}\sqrt{\overline{v}_L}\right) < 0$, ensuring that $\overline{m}_L$ is a maximum. Next, the irresolute L chooses $\underline{m}_L$ such that it satisfies

$$\max_{m_L}\{-m_L + \frac{m_L + m_P}{m_L + m_P + m_T}\underline{v}_L - c_L\}.$$

The optimum satisfies the first-order condition, $m_T\underline{v}_L/(m_P + m_T + m_L)^2 - 1 = 0$, yielding

$$\underline{m}_L = \sqrt{m_T}\sqrt{\underline{v}_L} - m_T - m_P. \tag{4.16}$$

Substituting $\underline{m}_L$ into the second partial with respect to m_L yields $-2/\left(\sqrt{m_T}\sqrt{\underline{v}_L}\right) < 0$, ensuring that $\underline{m}_L$ is a maximum.

Next, P cooperates for both low and high mobilization when $c_P < c_P^l$. To verify this, note that P cooperates after $\overline{m}_L$ when

$$\frac{\overline{m}_L + m_P}{\overline{m}_L + m_P + m_T}v_P - c_P\overline{m}_L > \frac{\overline{m}_L}{\overline{m}_L + m_T}v_P,$$

or when

$$c_P < \frac{v_P}{\overline{m}_L}\left(\frac{\overline{m}_L + m_P}{\overline{m}_L + m_P + m_T} - \frac{\overline{m}_L}{\overline{m}_L + m_T}\right) \equiv c_P^h. \tag{4.17}$$

Next, P cooperates after $\underline{m}_L$ when

$$\frac{\underline{m}_L + m_P}{\underline{m}_L + m_P + m_T}v_P - c_P\underline{m}_L > \frac{\underline{m}_L}{\underline{m}_L + m_T}v_P,$$

or when

$$c_P < \frac{v_P}{\underline{m}_L}\left(\frac{\underline{m}_L + m_P}{\underline{m}_L + m_P + m_T} - \frac{\underline{m}_L}{\underline{m}_L + m_T}\right) \equiv c_P^l. \tag{4.18}$$

Since $\underline{m}_L < \overline{m}_L$, as shown in Equations (4.15) and (4.16), it follows that $c_P^l < c_P^h$. Therefore, the lowest constraint binds, and P cooperates for both mobilization levels when $c_P < c_P^l$.

Next, consider what proposals L accepts after choosing m_L. The resolute type accepts iff x satisfies

$$x\overline{v}_L \geq \frac{\overline{m}_L + m_P}{\overline{m}_L + m_P + m_T}\overline{v}_L - c_L \Leftrightarrow x \geq 1 - \frac{c_L + \sqrt{m_T}\sqrt{\overline{v}_L}}{\overline{v}_L} \equiv \overline{x},$$

and the irresolute type accepts iff x satisfies

$$x\underline{v}_L \geq \frac{\underline{m}_L + m_P}{\underline{m}_L + m_P + m_T}\underline{v}_L - c_L \Leftrightarrow x \geq 1 - \frac{c_L + \sqrt{m_T}\sqrt{\underline{v}_L}}{\underline{v}_L} \equiv \underline{x}. \quad (4.19)$$

Note that since $\underline{v}_L < \overline{v}_L$, it follows that $\underline{x} < \overline{x}$.

The next step is to verify that T proposes $x^* = \overline{x}$ if $m_L^* = \overline{m}_L$ and $x^* = \underline{x}$ if $m_L^* = \underline{m}_L$, given its beliefs in the separating equilibrium. First, when $m^* = \overline{m}_L$, T proposes $x^* = \overline{x}$ rather than some $x' > \overline{x}$, since the latter will win L's acceptance but leave T strictly worse off than $\overline{x}$. It also will not propose some $x' < \overline{x}$, provoking rejection, because

$$1 - \overline{x} \geq \frac{m_T}{\overline{m}_L + m_P + m_T} - c_T$$

is guaranteed to be true as long as $c_L, c_T > 0$. By the same logic, T will also make no proposal greater than $\underline{x}$ if $m_L^* = \underline{m}_L$, and it will propose no less when

$$1 - \underline{x} \geq \frac{m_T}{\underline{m}_L + m_P + m_T} - c_T, \quad (4.20)$$

which again is guaranteed to be true since $c_L, c_T > 0$.

Finally, we verify that each type of L's prescribed action is incentive-compatible. The resolute type must choose $m_L^* = \overline{m}_L$ rather than deviate to $\underline{m}_L$, which convinces T that it is irresolute. If T believes L to be irresolute, then it offers terms that only the irresolute type will accept, forcing $\overline{v}_L$ to reject in its return to optimal play. Therefore, $\overline{v}_L$ reveals its type via high mobilization when

$$-\overline{m}_L + \overline{x}(\overline{v}_L) \geq -\underline{m}_L + \frac{\underline{m}_L + m_P}{\underline{m}_L + m_P + m_T}\overline{v}_L - c_L,$$

which is guaranteed to be true since $\underline{v}_L < \overline{v}_L$ and $c_L > 0$. The irresolute type must choose $m_L^* = \underline{m}_L$ rather than deviate to $\overline{m}_L$, which convinces T that it is resolute. Since $\underline{v}_L$ is sure to accept any proposal that the resolute type accepts, its payoff to mimicking $\overline{v}_L$ involves accepting a relatively favorable proposal, meaning that it chooses low mobilization when $-\underline{m}_L + \underline{x}(\underline{v}_L) \geq -\overline{m}_L + \overline{x}(\underline{v}_L)$, which is true when

$$m_T \geq \frac{c_L^2\left(\sqrt{\overline{v}_L} + \sqrt{\underline{v}_L}\right)^2}{\overline{v}_L\left(\sqrt{\overline{v}_L} - \sqrt{\underline{v}_L}\right)^2} \equiv \hat{m}_T.$$

Note that this constraint is the same as that supporting the separating equilibrium in Proposition 1, defined in Inequality (4.13), because P's presence in the military balance is "canceled out" by L's adjustment of its mobilization decision. Therefore, the separating equilibrium exists for the conditions stated above.

The semi-separating equilibrium occurs when $m_T < \hat{m}_T$. Strategies and beliefs are as follows. $\overline{v}_L$: choose $m_L^* = \overline{m}_L$ and accept iff $x \geq \overline{x}$. $\underline{v}_L$: with probability h, choose $m_L^* = \overline{m}_L$ and accept iff $x \geq \underline{x}_h$; with probability $1 - h$, $\underline{v}_L$ choose $m_L^* = \underline{m}_L$ and accept iff $x \geq \underline{x}$. P: cooperate for all m_L. T: believe $\phi' = \phi(h)$ if $m_L^* = \overline{m}_L$, then propose $x^* = \overline{x}$ with probability r and $x^* = \underline{x}_h$ with probability $1 - r$; believe $\phi' = 0$ if $m_L^* = \underline{m}_L$, and propose $x^* = \underline{x}$. As before, there are no out-of-equilibrium beliefs.

Mobilization levels $\underline{m}_L$ and $\overline{m}_L$ are as defined above, and the resolute type's disincentive to deviate to $\underline{m}_L$ remains the same, since the target would again believe L to be irresolute and make a proposal that the resolute type is sure to reject. Likewise, T's optimal response to observing $\underline{m}_L$ is identical to that derived above. What remains is to establish the range of proposals $x \geq \underline{x}_h$ that $\underline{v}_L$ accepts after bluffing, T's posterior belief $\phi' = \phi(h)$ in the event of observing high mobilization, $\underline{v}_L$'s probability of choosing $\overline{m}_L$ that renders T indifferent between risking war and buying off both types, and T's probability of risking war that renders $\underline{v}_L$ indifferent between bluffing and honest revelation.

First, after choosing high mobilization, the irresolute L has a new expected value for war, which determines the proposals it accepts in equilibrium. It accepts proposals that satisfy

$$x\underline{v}_L \geq \frac{\overline{m}_L + m_P}{\overline{m}_L + m_P + m_T}\underline{v}_L - c_L \Leftrightarrow x \geq 1 - \frac{\sqrt{m_T}}{\sqrt{\underline{v}_L}} - \frac{c_L}{\underline{v}_L} \equiv \underline{x}_h.$$

Since types differ only in their valuation for war, the irresolute L is sure to accept both this proposal and any more generous one, that is, $x \geq \overline{x}$, that the resolute type accepts. However, the resolute type will not accept $\underline{x}_h$, since its expected value for war is higher. This creates a risk-return trade-off for T if it observes high mobilization: it can propose $x^* = \underline{x}_h$, which $\underline{v}_L$ accepts and $\overline{v}_L$ rejects, or $x^* = \overline{x}$, which both accept. As in similar models, it is straightforward to show that as long as one type accepts a proposal, the costliness of war ensures that T never makes a proposal that both reject.

Second, let $\underline{v}_L$ choose $m^* = \overline{m}_L$ with probability h. Then, by Bayes' Rule, T's posterior belief that L is resolute, given $m^* = \overline{m}_L$, is $\Pr(\overline{v}_L|\overline{m}_L) = \phi/(\phi + h(1 - \phi))$. Third, $\underline{v}_L$ chooses the probability of bluffing, h, to render the target indifferent over proposing $\overline{x}$ and $\underline{x}_h$, after observing high mobilization. Formally, h solves

$$\frac{\phi}{\phi + h(1 - \phi)}\left(\frac{m_T}{\overline{m}_L + m_P + m_T} - c_T\right) + \left(1 - \frac{\phi}{\phi + h(1 - \phi)}\right)(1 - \underline{x}_h) = 1 - \overline{x},$$

where $\Pr(\bar{v}_L | \bar{m}_L)$ is as defined above. This yields its equilibrium probability of choosing high mobilization, b^*, where

$$b^* = \frac{\phi\,(c_L + c_T \bar{v}_L)\,\underline{v}_L}{(1 - \phi)c_L\,(\bar{v}_L - \underline{v}_L)}.$$

Note that b^* takes on plausible values, $0 \leq b^* \leq 1$, when types are well defined, or when $\underline{v}_L \leq (c_L \bar{v}_L\,(1 - \phi))\,/\,(c_L + \phi c_T \bar{v}_L)$. When types are well defined, T finds it profitable to potentially separate them with its proposal – a precondition for a genuine risk of war – so I maintain this assumption throughout the analysis.

Finally, rendered indifferent over its available proposals, T chooses a probability of risking war, r, that renders $\underline{v}_L$ indifferent over bluffing and honestly revealing its type. Thus, r must solve $-\underline{m}_L + \underline{x}\,(\underline{v}_L) = \bar{m}_L + r\,(\underline{x}_b \underline{v}_L) + (1 - r)\,(\bar{x}\underline{v}_L)$, yielding the equilibrium probability with which T makes the risky proposal,

$$r^* = 1 + \frac{\sqrt{m_T}\sqrt{\bar{v}_L}}{c_L}\left(1 - \frac{2}{1 + \sqrt{\underline{v}_L}/\sqrt{\bar{v}_L}}\right).$$

Algebra shows that T's mixing probability takes on plausible values $0 < r^* \leq 1$ when $m_T < \hat{m}_T$, as defined above, which establishes $\hat{m}_T$ as the cutpoint dividing the equilibrium space between pure separating and semi-separating equilibria.

It remains to show that pooling equilibria in which both player-types choose the same action do not exist. First, the resolute L's refusal to pool with the irresolute type, established above, is strictly true given $\underline{v}_L < \bar{v}_L$ and any out-of-equilibrium beliefs T might hold, since L is sure to receive its reservation value after high mobilization regardless of T's strategy. Therefore, there can exist no equilibrium in which both types pool on low mobilization (or in which the resolute L probabilistically chooses low mobilization). Second, the irresolute type is unwilling to pool on high mobilization, since by construction T's priors are sufficiently optimistic that it will risk war – that is, propose $\underline{v}_L$'s reservation value, which $\bar{v}_L$ goes on to reject – in a pooling equilibrium. For T to risk war, it must be the case that

$$\phi\left(\frac{m_T}{\bar{m}_L + m_P + m_T} - c_T\right) + (1 - \phi)(1 - \underline{x}_b) > 1 - \bar{x} \Leftrightarrow \phi < \frac{c_L\,(\bar{v}_L - \underline{v}_L)}{\bar{v}_L\,(c_L + c_T \underline{v}_L)} \equiv \phi_w.$$

To tolerate pooling under these conditions, the irresolute type would require $-\bar{m}_L + \underline{x}_b \underline{v}_L \geq -\underline{m}_L + \underline{x}\,(\underline{v}_L)$, which cannot be true as long as war is costly and $\underline{v}_L < \bar{v}_L$. Therefore, in the three-player game with a committed partner, there are two equilibria: (1) a fully separating equilibrium when $m_T \geq \hat{m}_T$, and (2) a semi-separating equilibrium when $m_T < \hat{m}_T$. Since L's mobilization decisions take P's contribution into account for both low and high mobilization, the equilibria and the threshold dividing them are the same as in the two-player game. □

Proof of Proposition 4.3. I first prove the existence of separating, semi-separating, and pooling equilibria, then show that equilibria in which players pool on high mobilization and in which the resolute L probabilistically chooses low mobilization do not exist.

The separating equilibrium exists when $c_P^l \leq c_P < c_P^h$ and $m_T \geq \max\{m_T^\dagger, \tilde{m}_T\}$. Strategies and beliefs are as follows. $\overline{v}_L$: choose $m_L^* = \overline{m}_L$ and accept iff $x \geq \overline{x}$. $\underline{v}_L$: choose $m_L^* = \underline{m}_L$ and accept iff $x \geq \underline{x}$. P: cooperate iff $m_L^* = \underline{m}_L$. T: believe $\phi' = 1$ if $m_L^* = \overline{m}_L$ and propose $x^* = \overline{x}$; believe $\phi' = 0$ if $m_L^* = \underline{m}_L$ and propose $x^* = \underline{x}$. Since each possible mobilization level is chosen, there are no out-of-equilibrium beliefs.

To prove the existence of this equilibrium, begin by deriving each type of L's optimal mobilization level given its type and P's strategy. Since P cooperates only if $m_L^* = \underline{m}_L$, the resolute L's optimal mobilization, $\overline{m}_L$, is the same as given above in Equation (4.8). The irresolute L expects P's support, so its optimum, $\underline{m}_L$, is the same as given in Equation (4.16).

Given mobilization levels, we can now derive the conditions that support P's strategy of cooperating if $m_L^* = \underline{m}_L$ and defecting if $m_L^* = \overline{m}_L$. Formally, this requires that both

$$\frac{\underline{m}_L + m_P}{\underline{m}_L + m_P + m_T} v_P - c_P \underline{m}_L \geq \frac{\underline{m}_L}{\underline{m}_L + m_T} v_P$$

and

$$\frac{\overline{m}_L}{\overline{m}_L + m_T} v_P > \frac{\overline{m}_L + m_P}{\overline{m}_L + m_P + m_T} v_P - c_P \overline{m}_L$$

be simultaneously true. This system of inequalities is true when $c_P^l \leq c_P < c_P^h$, where c_P^h and c_P^l are as defined in Inequalities (4.17) and (4.18), respectively.

Next, consider what proposals L accepts after choosing m_L. The resolute type accepts any $x \geq \overline{x}$, as defined by Inequality (4.9), since its payoffs are identical to the separating equilibrium in the two-player game. The irresolute type accepts any $x \geq \underline{x}$, as defined by Inequality (4.19), since its payoffs are identical to the separating equilibrium in the three-player game with a committed partner. Further, $\underline{x}$ is also the same proposal defined by Inequality (4.10), since L's mobilization decision "cancels out" the effect of m_P on the proposal. Thus, as before, $\underline{x} < \overline{x}$.

Next, we can verify that T proposes $x^* = \overline{x}$ if $m_L^* = \overline{m}_L$ and $x^* = \underline{x}$ if $m_L^* = \underline{m}_L$, given its beliefs defined above. First, when $m_L^* = \overline{m}_L$, T proposes $x^* = \overline{x}$ rather than some $x' \geq \overline{x}$, since it will win L's acceptance but leave T strictly worse off than $\overline{x}$. It also will not propose some $x' < \overline{x}$, provoking rejection, as long as $c_L, c_T > 0$, as defined in Inequality (4.11). By the same logic, T will also make no proposal greater than $\underline{x}$ if $m_L^* = \underline{m}_L$, nor will it propose any less, since $c_L, c_T > 0$ as shown for the identical decision in Inequality (4.20).

Finally, we verify that each player-type of L's prescribed action is incentive-compatible. First, the resolute type must choose $m_L^* = \overline{m}_L$ rather than deviate to $\underline{m}_L$, which convinces T that is irresolute and wins P's cooperation. If T believes it to be irresolute, then it offers terms that only the irresolute type will accept, forcing the irresolute type to reject in its return to optimal play. Therefore, the resolute L reveals its type via high mobilization when

$$-\overline{m}_L + \overline{x}(\overline{v}_L) \geq -\underline{m}_L + \frac{m_L + m_P}{\underline{m}_L + m_P + m_T}\overline{v}_L - c_L, \tag{4.21}$$

which is true when

$$m_T \geq \frac{m_P^2 \underline{v}_L}{\left(\sqrt{\overline{v}_L} - \sqrt{\underline{v}_L}\right)^4} \equiv m_T^\dagger. \tag{4.22}$$

Finally, the irresolute type must choose $m_L^* = \underline{m}_L$ rather than deviate to $\overline{m}_L$, which convinces T that it is resolute and ensures P's cooperation. Since it is sure to accept any proposal that the resolute type does, its payoff to mimicking the resolute type involves accepting a relatively favorable proposal, meaning that it chooses low mobilization when $-\underline{m}_L + \underline{x}(\underline{v}_L) \geq -\overline{m}_L + \overline{x}(\underline{v}_L)$, which is true when

$$m_T \geq \frac{\left((-c_L + m_P)\overline{v}_L + c_L\underline{v}_L\right)^2}{\overline{v}_L \left(\sqrt{\overline{v}_L} - \sqrt{\underline{v}_L}\right)^4} \equiv \tilde{m}_T. \tag{4.23}$$

Therefore, the separating equilibrium exists when $c_P^l \leq c_P < c_P^h$ and $m_T \geq \max\{\tilde{m}_T, m_T^\dagger\}$.

The semi-separating equilibrium exists when $c_P^l \leq c_P < c_P^h$ and $m_T^\dagger \leq m_T < \tilde{m}_T$. Strategies and beliefs are as follows. $\overline{v}_L$: choose $m_L^* = \overline{m}_L$ and accept iff $x \geq \overline{x}$. $\underline{v}_L$: with probability h, choose $m_L^* = \overline{m}_L$ and accept iff $x \geq \underline{x}_h$; with probability $1 - h$, $\underline{v}_L$ choose $m_L^* = \underline{m}_L$ and accept iff $x \geq \underline{x}$. P: cooperate iff $m_L^* = \underline{m}_L$. T: believe $\phi' = \phi(h)$ if $m_L^* = \overline{m}_L$, then propose $x^* = \overline{x}$ with probability r and $x^* = \underline{x}_h$ with probability $1 - r$; believe $\phi' = 0$ if $m_L^* = \underline{m}_L$, and propose $x^* = \underline{x}$. As before, there are no out-of-equilibrium beliefs.

Mobilization levels $\underline{m}_L$ and $\overline{m}_L$ are as defined in the separating equilibrium, and the condition supporting the resolute type's disincentive to deviate to $\underline{m}_L$ remains the same, $m_T \geq m_T^\dagger$, since the target would again believe L to be irresolute and make a proposal that the resolute type is sure to reject. Likewise, T's optimal response to observing $m_L^* = \underline{m}_L$ is identical to that derived for the separating equilibrium immediately above, as is the range of proposals the irresolute type accepts if it chooses high mobilization, that is, $\underline{x}_h$ from Inequality (4.14), and T's choice between buying off both types or risking war. What remains is to establish T's posterior belief $\phi' = \phi(h)$ in the event of high mobilization, $\underline{v}_L$'s probability of choosing $\overline{m}_L$ that renders T indifferent between risking war and buying off both types, and T's probability of risking war that renders $\underline{v}_L$ indifferent between bluffing and honest revelation.

The first step is to derive the probability with which $\underline{v}_L$ chooses $m_L^* = \overline{m}_L$, or h. By Bayes's Rule, T's posterior belief that L is resolute, given $m^* = \overline{m}_L$, is

$\Pr(\overline{v}_L | \overline{m}_L) = \phi / (\phi + h(1 - \phi))$. Third, $\underline{v}_L$ chooses the probability of bluffing, h, to render the target indifferent over proposing $\overline{x}$ and $\underline{x}_h$, after observing high mobilization. Formally, h solves

$$\frac{\phi}{\phi + h(1 - \phi)} \left(\frac{m_T}{\overline{m}_L + m_T} - c_T \right) + \left(1 - \frac{\phi}{\phi + h(1 - \phi)} \right) (1 - \underline{x}_h) = 1 - \overline{x},$$

where $\Pr(\overline{v}_L | \overline{m}_L)$ is as defined above. This yields its equilibrium probability of choosing high mobilization, h^*, where

$$h^* = \frac{\phi (c_L + c_T \overline{v}_L) \underline{v}_L}{(1 - \phi) c_L (\overline{v}_L - \underline{v}_L)}.$$

Note that h^* takes on plausible values, $0 \leq h^* \leq 1$, when types are well defined, or when $\underline{v}_L \leq (c_L \overline{v}_L (1 - \phi)) / (c_L + \phi c_T \overline{v}_L)$ as established above.

Finally, rendered indifferent over its available proposals, T chooses a probability of risking war, r, that renders $\underline{v}_L$ indifferent over bluffing and honestly revealing its type. Thus, r must solve $-\underline{m}_L + \underline{x}(\underline{v}_L) = \overline{m}_L + r(\underline{x}_h \underline{v}_L) + (1 - r)(\overline{x} \underline{v}_L)$, yielding the equilibrium probability with which T makes the risky proposal,

$$r^* = \frac{-m_P \overline{v}_L - \sqrt{m_T} \sqrt{\overline{v}_L} (\sqrt{\overline{v}_L} - \sqrt{\underline{v}_L})^2 + c_L (\overline{v}_L - \underline{v}_L)}{c_L (\overline{v}_L - \underline{v}_L)}$$

Algebra shows that T's mixing probability takes on plausible values when $m_T < \tilde{m}_T$. Therefore, the semi-separating equilibrium exists when $c_P^l \leq c_P < c_P^h$ and $m_T^\dagger \leq m_T < \tilde{m}_T$.

The pooling equilibrium in which both types choose low mobilization occurs when $m_T < \min\{m_T^\dagger, \hat{m}_T\}$. Strategies and beliefs are as follows. $\overline{v}_L$: choose $m_L^* = \underline{m}_L$ and accept iff $x \geq \overline{x}_l$; if $m_L^* = \overline{m}_L$, accept iff $x \geq \overline{x}$. $\underline{v}_L$: choose $m_L^* = \underline{m}_L$ and accept iff $x \geq \underline{x}$; if $m_L^* = \overline{m}_L$, accept iff $x \geq \underline{x}_h$. P: cooperate iff $m_L^* = \underline{m}_L$. T: believe (in equilibrium) $\phi' = \phi$ if $m_L^* = \underline{m}_L$, and propose $x^* = \underline{x}$; believe (out of equilibrium) $\phi' = 0$ if $m_L^* = \overline{m}_L$, then propose $x^* = \underline{x}$.

Mobilizations $\underline{m}_L$ and $\overline{m}_L$ are as defined above, and the proposals each type accepts – $x \geq \underline{x}$ for $\underline{v}_L$ if $m_L^* = \underline{m}_L$, $x \geq \underline{x}_h$ for $\underline{v}_L$ if $m_L^* = \overline{m}_L$, and $x \geq \overline{x}$ for $\overline{v}_L$ if $m_L^* = \overline{m}_L$ – are as defined in the separating and semi-separating equilibria. It remains to determine the proposals that the resolute L accepts following $m_L^* = \underline{m}_L$, to verify that T risks war following $m_L^* = \underline{m}_L$, and to identify conditions under which each type of L takes its prescribed actions.

First, after choosing low mobilization, the resolute L accepts proposals that satisfy

$$x \overline{v}_L \geq \frac{\underline{m}_L + m_P}{\underline{m}_L + m_P + m_T} \overline{v}_L - c_L \Leftrightarrow x \geq 1 - \frac{c_L}{\overline{v}_L} - \frac{\sqrt{m_T}}{\sqrt{\overline{v}_L}} \equiv \overline{x}_l. \tag{4.24}$$

Since types differ only in their valuation of the stakes, the irresolute L is sure to accept both this proposal and any more generous one, i.e. $x \geq \overline{x}_l$, that the resolute type accepts. However, the resolute type will not accept $\underline{x}$, since its expected value for war is higher. This creates a risk-return trade-off for T if it observes low mobilization: it can propose $x^* = \underline{x}$, which $\underline{v}_L$ accepts and $\overline{v}_L$ rejects, or $x^* = \overline{x}_l$, which both accept. As in similar models, it is straightforward to show that as long as one type accepts a proposal, the costliness of war ensures that T never makes a proposal that both types are sure to reject.

Next, we verify that on the equilibrium path T proposes $x^* = \underline{x}$ when $\phi < \phi_w$ after $m_L^* = \underline{m}_L$. As before, T will never concede any more than necessary to induce a type's acceptance, nor will it make a proposal that neither type accepts. Therefore, it proposes $x^* = \underline{x}$, generating a risk of war, when

$$\phi\left(\frac{m_T}{\underline{m}_L + m_P + m_T} - c_T\right) + (1 - \phi)(1 - \underline{x}) > 1 - \overline{x}_l \Leftrightarrow \phi < \frac{c_L\left(\overline{v}_L - \underline{v}_L\right)}{\overline{v}_L\left(c_L + c_T\underline{v}_L\right)} \equiv \phi_w,$$

as stated above.

Finally, we show that each type of L's strategy is incentive-compatible. Since both types choose $m_L^* = \underline{m}_L$ in equilibrium, we must first define T's out-of-equilibrium beliefs in the event that it observes high mobilization. As stated earlier, I assign beliefs $\phi' = 0$ in the event of high mobilization, that is, that T believes that only the *irresolute* type would deviate to high mobilization. While this seems counterintuitive, it is nonetheless the *only* belief that satisfies Cho and Kreps' (1987) Intuitive Criterion when $\underline{v}_L$ has an incentive to bluff, or when $m_T < \tilde{m}_T$, for two reasons. First, the resolute type never does better by deviating to high mobilization, regardless of T's strategy, because it can always guarantee itself its reservation value should it deviate from low mobilization; thus, if it prefers receiving its reservation value under low mobilization to high, then it cannot profit by deviating to high mobilization for *any* of T's possible strategies. Second, however, the irresolute type *can* do better than its equilibrium payoff for *one* of T's possible strategies – that is, if T will propose the resolute type's reservation price – regardless of whether that strategy will be played off the equilibrium path (see also Fudenberg and Tirole 1991, pp. 447–451). Therefore, should T observe high mobilization, it believes that only the irresolute type would have attempted this deviation and as such proposes its reservation value, $\underline{x}_h$, which the irresolute type accepts but which the resolute type rejects.

It is worth noting that since the irresolute type has no incentive to bluff about its resolve when $m_T \geq \tilde{m}_T$, any set of out-of-equilibrium beliefs and consistent strategies for T support the equilibrium. Thus, while the equilibrium is supported under a wider range of conditions when $m_T \geq \tilde{m}_T$, I characterize results based on out-of-equilibrium beliefs $\phi' = 0$, because $\underline{v}_L$ does weakly worse than the resolute type along any equilibrium path and can be considered the type with the most incentive to deviate even if the Intuitive Criterion does not strictly apply (Fudenberg and Tirole 1991, p. 447), as is the case when $m_T \geq \tilde{m}_T$.

Begin with the resolute type. When it chooses $m_L^* = \underline{m}_L$, T cannot update its beliefs and so proposes $x^* = \underline{x}$, which $\bar{v}_L$ rejects since $\underline{x} < \bar{x}_l$; should it deviate and choose high mobilization, T believes it to be irresolute and proposes terms, $\underline{x}_h$, that the resolute type is sure to reject (however, note that by rejecting, $\bar{v}_L$ gets in expectation exactly what it would if T believed L to be resolute and proposed $\bar{x}$). Therefore, the resolute L chooses low mobilization when

$$-\underline{m}_L + \frac{\underline{m}_L + m_P}{\underline{m}_L + m_P + m_T}\bar{v}_L - c_L \geq -\overline{m}_L + \frac{\overline{m}_L}{\overline{m}_L + m_T}\bar{v}_L - c_L,$$

or when $m_T \leq m_T^\dagger$, where $m_T^\dagger$ is as defined in Inequality (4.22). Next, consider the irresolute type. When it chooses $m_L^* = \underline{m}_L$, T proposes $x^* = \underline{x}$, which it accepts; should it deviate and choose high mobilization, T believes it to be irresolute, and it proposes $\underline{x}_h$, which $\underline{v}_L$ accepts. Therefore, the irresolute L chooses low mobilization when $-\underline{m}_L + \underline{x}\left(\underline{v}_L\right) \geq -\overline{m}_L + \underline{x}_h\left(\underline{v}_L\right)$, which is guaranteed to be true since $\underline{v}_L < \bar{v}_L$ and $c_L > 0$. However, note that the irresolute L *would* profit from a deviation if T's out-of-equilibrium beliefs were such that it proposed the resolute type's reservation value under high mobilization, since $-\overline{m}_L + \bar{x}\left(\underline{v}_L\right) > -\underline{m}_L + \underline{x}\left(\underline{v}_L\right)$ when $m_T < \tilde{m}_T$. Thus, since at least one of T's possible out-of-equilibrium strategies can be profitable for $\underline{v}_L$ – where the same is not true for the resolute type – T's belief $\phi' = 0$ satisfies the Intuitive Criterion. Therefore, the pooling equilibrium in which both types of L choose $m_L^* = \underline{m}_L$ exists when $c_P^l \leq c_P < c_P^h$ and $m_T < m_T^\dagger$.

It remains to show that there exists no pooling equilibrium in which both types of L choose high mobilization and no semi-separating equilibrium in which the resolute L probabilistically chooses low mobilization. First, consider a potential pooling equilibrium on high mobilization. The irresolute L must be willing to accept its minmax payoff following $m^* = \overline{m}_L$ if T chooses to make a proposal that only the irresolute type accepts – which T does by construction – rather than revealing itself as irresolute and receiving its minmax payoff under a lower mobilization level. Formally, this requires $-\overline{m}_L + \underline{x}_h\left(\underline{v}_L\right) \geq -\underline{m}_L + \underline{x}\left(\underline{v}_L\right)$, which cannot be true as long as war is costly and $\underline{v}_L < \bar{v}_L$. Therefore, such a pooling equilibrium cannot exist. Second, consider a semi-separating equilibrium in which the resolute L probabilistically chooses low mobilization. This requires that the resolute type can be rendered indifferent over revealing itself as resolute with high mobilization and choosing low mobilization, after which T randomizes its offers. However, in any equilibrium, the resolute type does no better than its minmax value (given the ultimatum bargaining protocol), so it cannot be rendered indifferent across the two actions, since T's probability of risking war or not has no effect on $\bar{v}_L$'s payoffs. Thus, there can be no semi-separating equilibrium in which the resolute L probabilistically masks its type. Therefore, in the three-player case with a skittish partner, there are three equilibria: (1) a fully separating equilibrium,

(2) a semi-separating equilibrium in which the resolute type bluffs probabilistically, and (3) a pooling equilibrium in which both types of L choose low mobilization. $\square$

Proof of Proposition 4.4. To verify this claim, I show first that the probability of war is greater under the two-player case when $m_T \geq m_T^\dagger$. The equilibrium probability of war in the two player case is zero when $m_T \geq \hat{m}_T$ (separating equilibrium) and $h_2^* r_2^* \phi$ when $m_T < \hat{m}_T$ (semi-separating equilibrium). In the three-player case, it is zero when $m_T \geq \tilde{m}_T$ (separating equilibrium) and $h_3^* r_3^* \phi$ when $m_T < \tilde{m}_T$ (semi-separating equilibrium). Simple algebra shows that $h_2^* r_2^* \phi > h_3^* r_3^* \phi$ as long as all variables hold plausible values. Further, $\tilde{m}_T < \hat{m}_T$ is also true, ensuring that the three-player probability of war occupies a smaller region of the parameter space, as long as

$$c_L > \frac{m_P}{2\left(1 - \underline{v}_L/\overline{v}_L\right)},$$

which must be satisfied if the semi-separating equilibrium under three players is to exist at all, because $\tilde{m}_T > 0$ only when

$$c_L > \frac{m_P}{\left(1 - \underline{v}_L/\overline{v}_L\right)},$$

which is more difficult to satisfy and therefore the binding constraint.

Finally, note that war occurs with probability ϕ when $m_T < m_T^\dagger$, but when $m_T < m_T^\dagger$ in the two-player case, the probability of war is either zero (separating equilibrium) or $h_2^* r_2^* \phi$ (semi-separating equilibrium), where $0 < h_2^* < 1$ and $0 < r_2^* < 1$, ensuring that both are less than ϕ. $\square$

Proof of Proposition 4.5. To verify this claim, it is sufficient to show that the equilibrium probability of war in the three-player case with a skittish partner decreases in m_T. First, for the lowest values of m_T such that $m_T < m_T^\dagger$, the probability of war is ϕ. Second, when $m_T^\dagger \leq m_T < \tilde{m}_T$, the probability of war is $h_3^* r_3^* \phi$, which itself decreases in m_T, or

$$\frac{\partial h_3^* r_3^* \phi}{\partial m_T} = \frac{\phi^2 \sqrt{\overline{v}_L}\left(c_L + c_T \overline{v}_L\right) \underline{v}_L}{-2\left(1 - \phi\right) c_L^2 \sqrt{m_T}\left(\sqrt{\overline{v}_L} + \sqrt{\underline{v}_L}\right)^2} < 0.$$

Finally, when $m_T \geq \tilde{m}_T$, the probability of war is zero. Therefore, since the probability of war never increases in m_T and since higher values of m_T are associated with lower probabilities of war, the probability of war weakly decreases in m_T in equilibrium. $\square$

Proof of Proposition 4.6. To verify this claim, it is sufficient to show that the threshold $m_T^\dagger$ below which the pooling equilibrium exists increases in m_P,

$$\frac{\partial m_T^\dagger}{\partial m_P} = \frac{2m_P \underline{v}_L}{\left(\sqrt{\overline{v}_L} - \sqrt{\underline{v}_L}\right)^4} > 0.$$

This ensures that the pooling equilibrium exists for a wider range of parameter values – specifically, for ever stronger targets – as m_P increases. $\qquad\square$

5

Durability, Balancing, and Conflict Expansion

> If we stayed neutral, whichever side won would surely punish Turkey for not
> having joined them, and would satisfy their territorial ambitions at our expense.
> Mehmed Talaat Pasha
> *Source Records of the Great War, Vol. II,*
> ed. Charles F. Horne, p. 398

Their military benefits aside, coalitions are aggregations of power that can
provoke serious fears about the future in third parties that observe them waging war against their targets. Just as with the formation of alliances, coalitions
may provoke counter-coalitions of their own (Vasquez 2009, ch. 5), and simple
verbal statements of reassurance or limited intent are unlikely to be convincing (Kydd 2005). For example, in 1979, the Red Army entered Afghanistan
to assist that country's Communist government against a popular insurgency,
hoping to prevent the "loss" of Afghanistan to the West and to salvage Soviet
prestige (Kalinovsky 2011, p. 24). Despite these ostensibly localized aims – to
say nothing of the Soviet Union's considerable military might – neighboring
Pakistan chose to support the Afghan rebels, fearing a subsequent invasion
if the Soviet-Afghan coalition were successful (Amin 2010, pp. 86–89).[1] On
the other hand, in 1990, another superpower, the United States, assembled a
coalition to reverse Iraq's annexation of Kuwait, drawing military contributions from regional powers like Saudi Arabia, Egypt, and Turkey. Yet despite
long-standing rivalries with nearby Sunni governments and a fraught recent
history with the United States, neighboring Iran remained neutral, apparently
confident that a coalitional victory in Iraq would not pose a serious threat to its

---

[1] See also Jones (2010, p. 21), as well as (Weinbaum 1991a,1991b).

security. Twelve years later, however, as the United States led a different coalition into Iraq, Iran did not remain on the sidelines, providing material support to elements of what would become a widespread anti-occupation insurgency.

In each conflict, a powerful state led a coalitional war in the proximity of a third-party state with what would seem to have been reasonable fears of future encroachment. However, while the United States built a coalition that did not provoke outsiders to come to Iraq's aid in 1991, the Soviet Union in the 1980s and the United States after 2003 saw the fruits of military cooperation diminished by third-party balancing – that is, behavior designed to prevent the coalition from winning today's conflict or simply to drain resources the coalition members might otherwise use in the future. When third parties intervene in coalitional conflicts, they can undermine the gains of military cooperation, particularly capability aggregation, by draining coalitional resources, lengthening wars, drawing in still more participants, and creating more destruction and casualties (Shirkey 2012, Slantchev 2004). Expanded wars are inefficient (see Gartner and Siverson 1996, Werner 2000): after the fact, joiners wish they could have ensured the attacker's limited aims and attackers wish they could have reassured those joiners sufficiently to avoid a war. Thus, coalitions – like individual states – face a problem of credibly committing not to take undue advantage of their collective power (cf. Ikenberry 2001), especially as their aggregated military might poses larger threats to outsiders than that of states acting alone. While Chapter 2 shows that coalitions are significantly more likely to provoke third parties to assist their targets than are states acting alone, balancing occurs in only about 20% of coalitional crises. Accounting for this variation is the goal of this chapter.

Assessing coalitional threats is necessarily quite different than assessing the future threats posed by singletons, as both future intentions and future power are unknown quantities. Third parties must concern themselves not only with a coalition's power but also its *durability*, since those that sustain cooperation beyond the current conflict pose a greater threat to outsiders than those that disband. To explore this dynamic, I analyze a model in which (a) a third party chooses to remain neutral or intervene on the side of the target of a coalitional conflict, and (b) the coalition, if it wins, chooses afterwards whether to disband or remain intact. While the third party is uncertain over a member's specific interest in remaining in the coalition – and thus whether the coalition poses a future threat – it can leverage coalition members' revealed foreign policy preferences to make judgments about the likelihood that they sustain cooperation into the future. In equilibrium, the distribution of preferences within coalitions interacts with their military power to affect third parties' balancing decisions. When coalitions are likely to defeat their targets, third parties are most likely to balance against them when their members share similar preferences. This occurs because coalitions with homogeneous preferences are less likely to disband on their own than those with diverse preferences. However, when coalitions are relatively weaker, balancing has a smaller impact on the outcome of the war, rendering the costs too difficult to recover against durable coalitions; thus, observers are more likely to balance against weak, diverse coalitions than against weak, homogeneous coalitions. Therefore, an increasing diversity of coalitional preferences can either encourage *or* discourage conflict expansion, depending on whether the coalition is relatively strong or weak.

This chapter recovers the expected relationships in an empirical model of balancing behavior in interstate crises from 1946–2001, and further tests the subsidiary results of the theoretical model over the duration of cooperation among coalition partners after they win a war. Thus, theory and evidence in this chapter show that coalitions play a unique role in the expansion of conflicts, one that differs from singletons, since the effect of their aggregate military power depends on the diversity of their preferences. Singletons, by contrast, are unconditionally more likely to provoke balancing as they grow more powerful. In addition to clarifying how coalitions differ from equally powerful singletons, the chapter also shows that the tendency to associate multilateralism unconditionally with the successful deterrence of balancing (see Nye 2002) – that is, with the successful localization of conflicts – is misplaced. While unilateralism may provoke opposition, multilateralism can do the same when powerful coalitions form around long-term similar foreign policy preferences. Finally, while balancing coalitions aimed at confronting threats to the balance of power have received substantial attention (Powell 1999, Walt 1987, Waltz 1979), the possibility of coalitions themselves provoking balancing has not, and this chapter seeks to rectify that imbalance, distinguishing coalitions that provoke balancing from those that do not.

5.1 THE THREAT OF EXPANSION

Most conflicts remain bilateral, but a few expand, drawing in additional states and becoming regional or global wars. Both world wars, in fact, began as bilateral conflicts that expanded to draw in the great powers; in that sense, every war that remains bilateral is a world war that failed to expand. To the extent that an attacker sees its opponent gaining military support from other states – that is, to the extent that it provokes balancing – this is an unfortunate turn of events. If the attacker would not have posed a future threat to the states balancing against it, then the additional costs of war paid by both the attacker and its new opponent are wasteful; however, as mentioned in Chapter 2, third parties often find it difficult to gauge the extent of future threats, a dilemma made all the more difficult because intervening in a war now can prevent an adverse shift in power by facilitating the attacker's defeat (Powell 1999, ch. 5).

For singletons, keeping a target isolated and a conflict localized is a problem of reassurance, of convincing third parties that the belligerent will not reward today's quiescence with future aggression (see Kydd 2005, Nye 2002, Voeten 2005). However, where singletons pose a problem only of future intent, belligerent coalitions, whose future threats depend on the maintenance of military cooperation after winning today's war, pose problems of both future intent *and* future power. Even if coalitions ultimately *will* disband, they will be unable to communicate that information credibly with costless signals of reassurance – in other words, through simple diplomacy – because of an underlying incentive to misrepresent intentions (cf. Fearon 1995, Kydd 2005). To see why, suppose that a partner intends to remain *in* the coalition regardless of a third party's decision. If this is true, it has an incentive to discourage the third party from intervening, because (a) the coalition stands a better chance of winning today's war if it does not provoke balancing and (b) future cooperative endeavors with other members of the coalition are more effective when resources are not wasted on an expanded war today. If the third party were to believe promises to disband, then the coalition would make such a promise regardless of whether it intended to disband or not. As a result, such a signal would carry no information, and the observer would be unable to condition its strategy on the coalition's statements; doing so would lead it to forego interventions that would be in its interest, opening it up to exploitation in the future.[2]

How do states achieve reassurance and keep their targets isolated? One possibility is diplomatic multilateralism. When a great power's opponents on the UNSC support its proposed military actions, small states that might otherwise balance against the great power take note. If even a state's largest rivals refuse to veto its actions, then surely today's war aims are neither too expansive nor worthy of opposition (cf. Thompson 2006, Voeten 2005). However, to the extent that securing an international organization's approval also facilitates coalition-building (Chapman 2011, ch. 5), mistrustful third parties may find their problems magnified: What if this coalition does not disband after victory, using its new position for future gains at third parties' expense?

Another potential answer is to forego the assistance of certain coalition partners altogether. In the Korean War, for example, the United States turned down what would have been superlatively enthusiastic military support from Chiang Kai-Shek's Nationalist government on Taiwan. Instead, the United States neutralized the island – interposing parts of its navy between Taiwan and the mainland – to ensure that success in the war could not be used to further the Chinese Nationalists' goal of retaking mainland China just years after their retirement from the civil war (Stueck 1995, pp. 194, 195).[3] Likewise, the United States worked hard to keep Israel out of the 1991 Gulf War, Iraq's ballistic missile campaign against Israeli cities notwithstanding, in order to reassure nearby states of its intent to limit its aims to the restoration of Kuwaiti sovereignty (Atkinson 1993, pp. 92–93). Some contributions, consistent with

---

[2] The reader will note that the problem of reassurance is akin to the problem of signaling resolve in Chapter 4, although in this case the problem is convincing opponents of *benign* intent.

[3] The fear was not limited to provoking Chinese intervention; American decision makers believed that threatening China might also provoke Soviet entry into the war.

Chapters 3 and 4, simply come at too great a cost: the potential expansion of the war.

By turning down contributions from partners that would most contribute to the reassurance problem, coalition leaders are left with a collection of states that are more likely to disband and, thus, less likely to pose a future threat to third parties. Returning to the Korean War example, the United States did work hard to bring Asian states into the coalition to signal non-imperialist aims and discourage balancing (Stueck 1995, pp. 56, 58), but this choice was as much about achieving something akin to diplomatic multilateralism as anything else. Nonetheless, to the extent that the states in a coalition have diverse preferences, they should be less likely to agree on taking advantage of an eventual victory to coerce third-party states, and statements about benign intent should more effectively achieve reassurance than those that emanate from a homogeneous coalition.

Partner choice, however, is not always a feasible solution; coalitions are built to win, and the primary goal of military victory can lead to some substantial compromises in terms of the costs paid to secure military cooperation (Riker 1962, Starr 1972). Geography, in particular, makes some states almost inevitable partners, like Kuwait for the United States in the 2003 Iraq War, and South Korea during the Korean War. In these cases, lead states cannot choose the partners with which they act, and the consequence is the difficulties associated with the task of reassurance. The exigencies of winning mean that geographical and military factors often produce coalitions that, all else being equal, lead states would like to avoid – coalitions that provoke balancing. To the extent that lead states build coalitions with a diverse set of partners, promises to disband should be more credible than would be the case if a more homogeneous set of partners were chosen, and in the following section I explore this strategic problem in the book's most complex model, which incorporates coalition formation, war expansion, and coalitional durability.

5.2 A THEORY OF FORMATION, BALANCING, AND DURABILITY

While Chapter 4 explored how intra-coalitional politics shapes the escalation of crises to war, this chapter's theoretical model turns to its effects on their expansion – specifically, on the decision of a third-party state (B) that may intervene on the side of the target or remain neutral. Like the previous chapters, the game begins with L's choice of forming a coalition with P, but in addition to weighing the costs of securing a partner's cooperation, it must also consider the risks that the third party will balance against it in today's war and, in the event of victory, that its partner will continue to cooperate with it. Just as third parties do in other models of balancing and alignment (Altfeld and Bueno de Mesquita 1979, Powell 1999), B might join the war to try to defeat L, possibly in combination with L's coalition partner, to eliminate a future threat today, but doing so is costly; as such, the third party would like to save the costs of joining the war if downstream threats will diminish on their own – that is, if T will win the war without assistance or if a victorious coalition of L and P will ultimately disband.

As in previous chapters, L pays a premium in return for military cooperation, but coalition-building may entail another cost in the form of an expansion of the conflict. To understand how the lead state approaches this trade-off, analysis of the model focuses on two primary factors: (a) L's military strength as a singleton or coalition leader and (b) the distribution of revealed preferences between L and its potential partner. While singletons provoke balancing under what we might call typical conditions – that is, when they are sufficiently likely to defeat the target on their own – the effect of the distribution of power for coalitions depends on the diversity of preferences within them. The more likely is the coalition to win today's conflict, the more a third party's intervention can affect the coalition's military fortunes, encouraging balancing under some conditions in order to prevent the coalition from turning against it in the future; a defeated coalition, after all, poses less of a future threat than a victorious one. Next, coalition members' revealed preferences – that is, their history of cooperation and diplomatic behavior – helps the observer state draw inferences about the coalition's durability, even as it remains uncertain over the partner's specific willingness to cooperate with the leader after today's conflict. In equilibrium, power and preferences interact to produce two distinct patterns of balancing behavior. When coalitions are relatively likely to defeat their targets, they are most likely to provoke balancing when they have homogeneous preferences, because third parties expect diverse coalitions to disband on their own, encouraging them to save the costs of intervention. However, when coalitions are relatively weak, a more diverse distribution of preferences provokes balancing, because observers hope to sap sufficient coalitional resources to encourage their subsequent breakup.

The model is similar to other treatments of alignment and war expansion in that costly intervention can influence both conflict outcomes (Altfeld and Bueno de Mesquita 1979, Werner 2000, Yuen 2009) and the terms on which an observer confronts potentially threatening states in the future (Powell 1999). However, my model sacrifices a richer set of alignment options, such as the option to "bandwagon," to focus more attention on the process of alignment decisions against coalitions: coalitions' military power, uncertainty over their durability, and how alignment decisions endogenously affect that durability.[4] Finally, where Wolford's (2014a) model of expansion and durability assumes that a coalition already exists between L and P, the expanded variant analyzed here offers L an initial choice of building a coalition or not. Since coalitions

---

[4] How coalitions affect incentives to bandwagon is a potentially interesting avenue for further research, and it is not without precedent: arguably, Russia's decision to pressure Serbia into acquiescence in the Kosovo War was designed to help it ensure a measure of influence in the region after a virtually inevitable NATO victory.

form endogenously in the model, comparisons across singletons and coalitions in the rate of balancing take that process – whereby L may avoid those partners most likely to provoke balancing – into account, minimizing concerns that selection bias may taint inferences drawn about the observed differences between the two.

5.2.1 The Model

Suppose that a lead state L is engaged in a conflict with a target state T, and it has the option of securing (at some cost) a partner P's cooperation, which increases L's chances of prevailing militarily. Whether L acts as a singleton or as part of a coalition, a third-party state B chooses whether or not to join the conflict on the side of the target – that is, to balance against L. While B most prefers that L lose the war, because it expects to engage in some kind of future bargaining with L where the outcome is influenced by relative power (cf. Powell 1999), it also prefers to face L alone in the future as opposed to an intact coalition, because winning a war can place the victors in an improved bargaining position. Imagine, for example, British fears if Germany successfully conquered Belgium and France in 1914, or Chinese fears in the face of a potential American-led reunification of the Korean peninsula in 1951. As such, after winning a coalitional war, P can remain in the coalition, continuing military cooperation over other issues – specifically, issues that affect B's well-being – from a position of strength, or leave. However, B is uncertain over what choice P will make, or just how durable the coalition will be. B thus makes its alignment decision with some uncertainty as to whether the coalition will remain intact to pose a future threat if it successfully defeats its target in today's conflict. This is the third party's basic dilemma when confronting a coalition: intervention today may facilitate L's defeat and influence P's decision over remaining in the coalition, yet fighting a war is wasteful if P would have left the coalition anyway had B remained neutral.

As shown in Figure 5.1, Nature begins the game by drawing two quantities, c_P and c_B, from the respective uniform distributions $c_P \sim U(0, \overline{c}_P)$ and $c_B \sim U(0, \overline{c}_B)$. First, c_P denotes P's costs for remaining in the coalition after today's war, and it is revealed only to P. These costs represent any disutility associated with cooperation, from resources invested to policy compromises (see Chapters 3 and 4), independent of the distributive outcome of future interactions with the third party. The higher these costs, the less willing is the partner, all else being equal, to remain in the coalition after today's conflict. Both L and B know only the distribution from which the partner's costs are drawn, ensuring that both are uncertain over whether the partner will remain in the coalition after today's war.[5] The second quantity, c_B, is the third party's costs for intervening

---

[5] I make this assumption to ensure that L's initial choice over taking on a partner is not also a signal to B of the likely durability of the coalition, which is substantively attractive in that states often find it difficult to predict future alignments before a war breaks out.

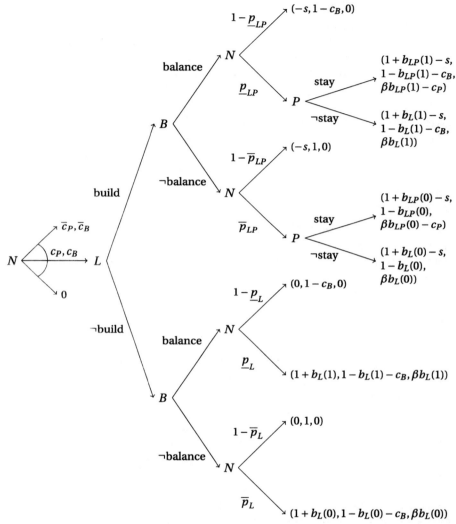

FIGURE 5.1. Coalition-building, conflict expansion, and coalitional durability

in the war as a belligerent, and likewise Nature reveals these costs only to B. This ensures that in the game's first strategic move, the lead state has only an imperfect forecast of both the probability of balancing and the postwar durability of a coalition.

Following Nature's move, L chooses whether or not to build a coalition, which requires that it pay a premium $s \geq 0$ in order to secure P's cooperation in the looming war.[6] Next, B chooses whether to remain neutral or to intervene on the target's side. Joining the war has two direct effects on B's payoffs. First, regardless of the outcome of the war, it must pay the direct costs of intervention, $c_B > 0$. Second, though costly, intervening may also decrease the other side's probability of victory. If it fights alone, L wins a war against T with probability $\overline{p}_L \in [\underline{p}_L, 1)$ if B remains neutral, but it wins with $\underline{p}_L \in (0, \overline{p}_L]$ should B side with the target, such that $\overline{p}_L \geq \underline{p}_L$. As in previous chapters, I assume that a given side's chances of success depend on the military endowments of each belligerent, or that L's probability of victory is

$$\overline{p}_L(m_L, m_T) \equiv \frac{m_L}{m_L + m_T} \quad \text{or} \quad \underline{p}_L(m_L, m_T, m_B) \equiv \frac{m_L}{m_L + m_T + m_B}.$$

Likewise, if L builds a coalition, it wins the war against the target with probability $\overline{p}_{LP} \in [\underline{p}_{LP}, 1)$ when the observer remains neutral and probability $\underline{p}_{LP} \in (0, \overline{p}_{LP}]$ should the third party enter the war, where $\overline{p}_{LP} \geq \underline{p}_{LP}$. B's intervention thus weakly improves the chances that the coalition loses today's war. Again, L's chances of victory given B's neutrality and intervention are

$$\overline{p}_{LP}(m_L, m_P, m_T) \equiv \frac{m_L + m_P}{m_L + m_P + m_T}$$

and

$$\underline{p}_{LP}(m_L, m_P, m_T, m_B) \equiv \frac{m_L + m_P}{m_L + m_P + m_T + m_B}.$$

Throughout, I abuse notation and simply write the reduced-form probabilities of victory as $\overline{p}_{LP}$, $\underline{p}_{LP}$, $\overline{p}_L$, and $\underline{p}_L$.

Should a coalition win today's conflict, P must then decide whether to stay in or leave the coalition. If the partner stays in the coalition, it pays the costs of cooperation (c_P) and receives a payoff for cooperating with the lead state in future bargaining with B, weighted by the extent to which its preferences tend to align with L's, or $\beta \in [0, 1]$. While coalition formation indicates that L and P are in sufficient agreement to fight T together today, β captures the extent to which they agree or disagree about B in the future: as β approaches one, the partner increasingly shares the lead state's preferences, and as it approaches zero, P's preferences are increasingly opposed to the lead state. Since it is common knowledge, β is an indicator of revealed preferences that the third

---

[6] Rather than model P's acceptance decision, as I do in Chapters 3 and 4, I abstract away from it here in order keep any complexities unrelated to P's final decision to stay or leave to a minimum; this also ensures that the act of joining does not send a signal over expected durability, keeping the game simple and focused on the more important decisions of balancing and durability. I have analyzed a fuller version of the game in which P accepts or rejects L's proposal before its costs for staying in the coalition are drawn (in this case, after the war), but the key results do not change in any meaningful way; as such, I keep the (relatively) simpler specification shown in Figure 5.1.

party can use, in addition to its beliefs over the partner's costs of remaining in the coalition for future interactions, to inform judgments over the coalition's durability.

To characterize postwar bargaining outcomes, I assume that there is some postwar game Γ during which B divides peacefully a good of unit size with L, possibly in cooperation with P, that reflects each side's relative military endowments (cf. Powell 1999). Additionally, remaining military endowments may be shaped by losses in capabilities drained by the initial conflict.[7] Therefore, if P stays in the coalition, P's payoff for the subsequent bargain with B is

$$b_{LP}(I) \equiv \frac{\alpha^I m_L + \alpha^I m_P}{\alpha^I m_L + \alpha^I m_P + \delta^I m_B}, \tag{5.1}$$

where $\alpha \in (0, 1)$ and $\delta \in (0, 1)$ are scalars that measure the shares of military capabilities remaining after the war as a result of B's intervention, which occurs when $I = 1$. On the other hand, $I = 0$ if B has not intervened and sapped its future opponents' resources.[8] Thus, for B, $(1 - \delta)$ represents the opportunity costs of intervention. If P leaves the coalition, then the deal struck between L and B does not reflect P's military capabilities, such that the outcome $b_L(I)$ is identical to $b_{LP}(I)$ as defined in Equation (5.1) but for the absence of P's capabilities m_P.

Just as it did in Chapter 3, P weighs the outcome of the crisis by the extent to which its preferences align with L's, allowing us to characterize its payoff function as

$$u_P = \begin{cases} \beta b_L(I) & \text{if leave} \\ \beta b_{LP}(I) - c_P & \text{if stay} \\ \beta b_L(I) & \text{if } \neg\text{build and } T \text{ loses} \\ 0 & \text{if } T \text{ wins,} \end{cases}$$

where P also has an interest the outcome of future bargaining with B if L does not build a coalition in the first place. However, in contrast to Chapter 3, there is no private component of the stakes of the crisis; here I am concerned only with P's preferences over the content of the distributive outcome.

Next, B's payoffs depend on the outcome of the first conflict, whether the coalition disbands, and the costs of intervention – all of which shape the terms of the bargain it expects to strike with L. In addition to increasing the coalition's chances of defeat, intervention also causes the leader and partner to expend

---

[7] Absent bargaining frictions, this is the division we would expect the parties to strike in lieu of war. The term, as shown next, reflects each side's expected value for fighting, though it saves the costs, allowing for an accurate reflection of the consequences of bargaining without introducing additional complications to the model. For a similar use of this solution as a baseline peace agreement, see Debs and Goemans (2010).

[8] Since I enters the expression as an exponent, both α and δ are equal to 1 when $I = 0$, and they are equal to α and δ, respectively, when $I = 1$.

extra resources, leaving them less able to threaten the observer after today's conflict. Should the coalition lose, its members pose the smallest subsequent threat, and B receives 1, or the whole of the good it would otherwise have to divide with L. Should the coalition win, its members pose a greater threat to the observer together than they do separately, since P's resources will be turned against B in addition to L's. We can now summarize B's payoff function. First, if the third party remains neutral, or $I = 0$,

$$u_B(\neg\text{balance}) = \begin{cases} 1 - b_{LP}(0) & \text{if } L \text{ builds, } T \text{ loses, and } P \text{ stays} \\ 1 - b_L(0) & \text{if } L \text{ builds, } T \text{ loses, and } P \text{ leaves} \\ 1 - b_L(0) & \text{if } L \neg\text{build and } T \text{ loses} \\ 1 & \text{if } T \text{ wins} \end{cases} \quad (5.2)$$

Second, if the third party intervenes, or $I = 1$,

$$u_B(\text{balance}) = \begin{cases} 1 - b_{LP}(1) - c_B & \text{if } L \text{ builds, } T \text{ loses, and } P \text{ stays} \\ 1 - b_L(1) - c_B & \text{if } L \text{ builds, } T \text{ loses, and } P \text{ leaves} \\ 1 - b_L(1) - c_B & \text{if } L \neg\text{build and } T \text{ loses} \\ 1 - c_B & \text{if } T \text{ wins} \end{cases}$$

$$(5.3)$$

Finally, the lead state's payoffs depend on any costs paid to build a coalition, the outcome of the first conflict, and subsequent bargaining between itself, possibly with P's assistance, and B. If it wins today's war, it receives 1 and survives into the future, but should it lose, it receives zero and does not bargain with B in the future. Assuming that the costs of the current war are sunk, L's payoff function is

$$u_L = \begin{cases} -s & \text{if } L \text{ builds and loses} \\ -s + 1 + b_{LP}(I) & \text{If } L \text{ builds and wins, } P \text{ stays} \\ -s + 1 + b_L(I) & \text{If } L \text{ builds and wins, } P \text{ leaves} \\ 0 & \text{if } L \neg\text{build and loses} \\ 1 + b_L(I) & \text{if } L \neg\text{build and wins,} \end{cases} \quad (5.4)$$

where I once again indicates whether B chooses to balance against L ($I = 1$) in today's conflict or remain neutral ($I = 0$).

Equations (5.1)–(5.4) highlight the strategic problems faced by each player in the game. First, L would like to bolster its chances of success in the current conflict, but doing so comes at the certain cost of securing cooperation up front and the uncertain costs of both provoking balancing and coalitional breakdown in the future. For B's part, intervening saps coalition members' resources even when it does not lead to their defeat, but it is also costly and reduces B's own future bargaining power. Further, B's best outcome in the event that the coalition wins still requires that it pay the costs of intervention, when the possibility exists that P might have left the coalition of its own volition – raising the specter of wasteful intervention that carries ex post regret. Finally, P would like to save the costs of cooperation in the future, but it cannot commit not to remain in the coalition if staying would be in its interest, which might provoke otherwise avoidable interventions. I turn in the following section to an analysis of how these incentives interact to influence balancing decisions, coalitional durability, and the formation of coalitions in equilibrium. While the third party's alignment decision resembles more traditional treatments in some respects (Altfeld and Bueno de Mesquita 1979, Powell 1999), I show next that introducing coalitional formation and durability into the process can change our understanding of the link between power, threat, and balancing.

5.2.2 Analysis

In this section, I answer two primary questions. First, when do third parties balance against coalitions by embarking on costly expansions of ongoing conflicts? Second, how do the conditions for balancing differ across coalitions and singletons? The answer depends, in both cases, on the distribution of preferences within coalitions, because it plays a key role in shaping coalitional durability and the specific incentives that third parties have at different distributions of power to intervene in the conflict. To establish a baseline for comparison, I first characterize balancing behavior in those cases in which L acts as a singleton, showing that more powerful states are more likely to provoke balancing than are weaker states. Then, I consider B's incentives to balance against a coalition, which depend on an interaction between the diversity of coalitional preferences and the lead state's chances of defeating the target if the third party does not intervene. In equilibrium, preference diversity discourages balancing against powerful coalitions but encourages it against weaker ones, and this interactive relationship is driven by the link between preference diversity and coalitional durability. To establish these claims, I characterize the model's unique Perfect Bayesian Equilibrium in full, then split the discussion into singleton and coalition behavior.[9]

Proposition 5.1 characterizes the strategies that constitute the PBE, which rest on two assumptions. Specifically, the third party's intervention must sap enough of the coalition's resources (that is, α must be sufficiently low) while *not* costing B too many of its own capabilities (δ must be sufficiently high). These two conditions ensure that balancing is worthwhile in principle – that is, that B's choice is substantively meaningful. Otherwise, if intervention were to fatally undermine B's own postwar bargaining position, or if it had no material impact on its future opponents' positions, then intervention is trivially

---

[9] Since (a) uninformed players move before informed players and (b) B's type cannot affect P's payoffs in the last move, posterior and out-of-equilibrium beliefs are trivial. As such, I omit discussion of them from the main text.

impossible in equilibrium. Therefore, the subsequent discussion relies on the twin restrictions that $\alpha < \alpha^\dagger$ and $\delta > \delta^\dagger$, ensuring that balancing is a viable military option for the potential third-party intervener.

Proposition 5.1. *For $\alpha < \alpha^\dagger$ and $\delta > \delta^\dagger$, the following strategies are part of the unique Perfect Bayesian Equilibrium. L builds a coalition iff $s < s^\dagger$ and acts unilaterally otherwise. If L acts as a singleton, B balances iff $c_B < \tilde{c}_B$ and remains neutral otherwise; if L builds a coalition, B balances iff $c_B < \hat{c}_B$ and remains neutral otherwise. If B balances, P stays in the coalition iff $c_P < \hat{c}_P$ and leaves otherwise; if B remains neutral, P stays in the coalition if $c_P < \underline{c}_P$ and leaves otherwise.*

In equilibrium, the lead state builds a coalition when $s < s^\dagger$, or when the costs of compensating its potential partner are not too high relative to the twin risks that B chooses to balance against the coalition – especially as against the risk of provoking balancing as a singleton – and that P will leave the coalition after the initial conflict. As discussed later, L may turn down potential coalition partners that would create an unacceptable risk of balancing, although the temptation of securing victory today may lead it to accept partners that, under different circumstances, it might otherwise avoid (Starr 1972). The third party, for its part, weighs the risk of staying out of the war, which allows a victorious L to face an isolated third party in the future, against paying the costs of intervention in order to confront L alongside its current target, which allows B to fight at a relatively favorable distribution of power. When it faces a coalition, however, the third party must also consider the probability that the coalition will disband and pose a diminished threat, as well as its ability to sap enough coalitional resources to facilitate the coalition's dissolution by intervening. As such, it balances against the lead state's side in the conflict when the upfront costs of intervention are sufficiently low. Finally, after the war, P weighs its ability to contribute militarily to bargaining against the third party against the costs of staying in the coalition and its own preference alignment with L. The result is that the coalition remains intact when the partner's costs for cooperation are sufficiently low.

To characterize the relationship between the model's parameters – in particular, military power and coalitional preference diversity – and balancing decisions, I begin the discussion with a characterization of B's choice if L acts as a singleton. This establishes a baseline, consonant with the extant literature on alignment and war expansion, against which to judge the results of the next section, where B chooses whether to balance against a coalition and where the coalition either endures or disbands after the war. I conclude the section with a discussion of L's choice over building a coalition or not, then discuss its implications for war expansion in general and the selection process shaping the data used in the empirical models of the subsequent section.

Singletons and Balancing

If the lead state chooses to act as a singleton, the third party's balancing problem is straightforward. Fighting today is costly, although it both drains L's resources and increases the chances that the target wins the war; remaining neutral, on the other hand, saves the costs of war but, by foregoing the chance to add T's capabilities to its own in a bid to eliminate a future threat, allows power effectively to shift against B for the future. As such, B intervenes when

$$\underline{p}_L \left(1 - b_L(1)\right) + \left(1 - \underline{p}_L\right) - c_B > \overline{p}_L \left(1 - b_L(0)\right) + \left(1 - \overline{p}_L\right),$$

or when the upfront costs of intervention are not too great:

$$c_B < \overline{p}_L b_L(0) - \underline{p}_L b_L(1) \equiv \tilde{c}_B. \tag{5.5}$$

The logic behind Inequality (5.5) mirrors that of more general treatments of choosing sides during war (Altfeld and Bueno de Mesquita 1979, Powell 1999); as the difference between L's chances of prevailing across balancing and neutrality, or $\overline{p}_L - \underline{p}_L$, grows, then intervening is able to recoup ever larger costs of fighting. In fact, as stated in Proposition 5.2, B is increasingly likely to intervene against a singleton L the more likely the lead state is to defeat its target in today's war.

Proposition 5.2. *The probability that B is of a type that balances against a singleton, $\Pr(c_B < \tilde{c}_B)$, increases in L's probability of winning today's war if B remains neutral, $\overline{p}_L$.*

Taking advantage of the fact that B's type is uncertain but drawn from a known uniform distribution, the probability of balancing is simply the probability that c_B falls below $\hat{c}_B$. Formally, this is $\Pr(c_B < \hat{c}_B) = \hat{c}_B/\overline{c}_B$, which is the same belief that the lead state holds over the probability of balancing in equilibrium. Since this probability increases as L's probability of winning absent balancing (p_L) increases, we can say that singletons with a greater chance of defeating their opponents should be more attractive targets for balancing than singletons that are weaker relative to their targets, because B's intervention has a much smaller impact on the lead state's chances of victory when its prospects are already poor. Conversely, intervening against a stronger lead state has a potentially larger impact on the future distribution of power.

We can view B's incentives to balance against L as a specific instance of the more general phenomenon of commitment problems generated by shifting power. To see this, return to Equation (5.2) and note that $\overline{p}_L b_L(0)$ is L's probability of victory against T if the third party remains neutral, weighted by the deal that B expects to strike with L if it wins the war; likewise, $\underline{p}_L b_L(1)$ is L's probability of defeating T if B intervenes, again weighted by the terms of the subsequent deal. As the difference between the first and second terms grows, the third party experiences an ever greater shift of relative power against itself during the course of today's war if it stays on the sidelines. As such, Equation (5.2) shows that when the shift in power is larger than the costs of fighting (cf. Powell 2004), the third party chooses to fight, as the consequences of future weakness are worse than the costs of conflict in the present. The logic behind the third party's strategy in this case is simple: fighting today may prevent an adverse shift in power resulting from L's victory in the war, but the costs of fighting may prove wasteful if L would have gone on to lose the war anyway. However, a fearful third party's calculus changes when it faces a coalition, which may not only lose today's war but, even if it wins, may disband before it can pose a threat in the future.

Coalitions and Balancing

If the third party's problem when facing a singleton is relatively simple, how does it judge the threat posed by – and, by extension, the attractiveness of balancing against – a coalition? Since B is uncertain over the costs the partner must pay to continue military cooperation with the lead state (c_P), it makes a guess about the durability of the coalition based on both its prior beliefs over c_P *and* the extent of revealed preference similarity between lead state and partner. Before characterizing B's decision, however, note that the partner remains in the coalition after victory when the net benefits of continuing military cooperation exceed the upfront costs of remaining in the coalition, or when

$$c_P < \beta \left(b_{LP}(I) - b_L(I) \right). \tag{5.6}$$

This yields two cutpoints over the partner's costs of remaining in the coalition that define when it stays and when it leaves, depending on whether the third party balanced or remained neutral in the crisis. Specifically, the partner remains in the coalition following intervention when

$$c_P < \beta \left(b_{LP}(1) - b_L(1) \right) \equiv \underline{c}_P,$$

and it stays if B remains neutral when

$$c_P < \beta \left(b_{LP}(0) - b_L(0) \right) \equiv \hat{c}_P.$$

In each case, P is more willing to remain in the coalition when its participation makes a larger difference in the deal extracted from B, but as in previous chapters, the similarity of revealed foreign policy preferences inside the coalition also affects the ease with which Inequality (5.6) can be satisfied. The more that the preferences of lead state and partner align – that is, as β increases – the higher the costs of cooperation the partner is willing to pay to maintain military cooperation with the lead state, because it is willing to pay a larger premium to increase the chances of receiving a more favorable outcome. In other words, relatively homogeneous coalitions are more likely to remain intact, to be durable, than relatively diverse coalitions.

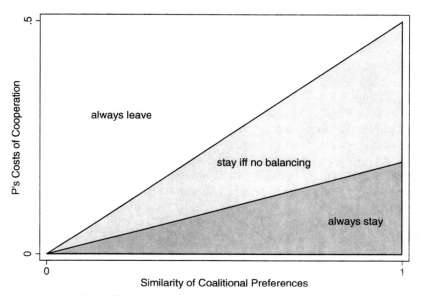

FIGURE 5.2. *P's willingness to stay in the coalition by the similarity of preferences*

Proposition 5.3. *For any I, the probability that P is of a type that stays in the coalition,* $\Pr(c_P < \underline{c}_P)$ *or* $\Pr(c_P < \hat{c}_P)$, *increases as* β *increases.*

Next, since the value of a coalition partner in extracting a better bargain from the third party is greater if the coalition's strength has not been degraded by fighting an expanded war beforehand, it must also be the case that $\underline{c}_P < \hat{c}_P$.[10] Thus, there are values of c_P for which the partner remains in the coalition if the third party remains neutral but leaves if the third party intervenes. In other words, P is more willing to stay in the coalition if the third party remains neutral, because the coalition will be left with more resources to bring to bear in the future, making it easier for the benefits of staying to offset the costs. Figure 5.2 shows that there are three ranges of partner types defined by their behavior in response to the third party's alignment choice: (a) always staying in the coalition when $c_P < \underline{c}_P$, (b) staying if the third party remains neutral and leaving after intervention when $\underline{c}_P \leq c_P < \hat{c}_P$, and (c) always leaving when $c_P \geq \hat{c}_P$.[11]

Figure 5.2 shows that both constraints ensuring the partner's willingness to remain in the coalition become easier to satisfy, such that more player-types of P are willing to stay, as coalition members' revealed preferences become increasingly similar. Specifically, the more the partner agrees with the leader

[10] Recall from Proposition 5.1 that this is ensured by $\alpha < \alpha^\dagger$ and $\delta > \delta^\dagger$.
[11] Parameter values in Figure 5.2 are fixed at $b_{LP}(1) = 0.65$, $b_L(1) = 0.45$, $b_{LP}(0) = 0.75$, and $b_L(0) = 0.25$.

over the need to extract concessions from B, the greater the costs it is willing to pay to cooperate in the future. However, since the partner's payoff for staying in the coalition is diminished after B's intervention drains some of the coalition's coercive resources, $\underline{c}_P$ increases more slowly in β than does $\hat{c}_P$. In other words, increasingly similar preferences over lower payoffs have less of an impact on P's decision than increasingly similar preferences over higher payoffs; if, for example, P's and L's remaining resources are so small as to render their combination nearly meaningless against B, then the costs of staying in the coalition are more difficult to justify. Thus, the partner is increasingly willing to condition its decision on the third party's alignment choice as β increases and coalitional preferences grow more similar, such that the coalition remains intact only if the third party remains neutral. This, of course, tempts the third party with a chance to split the coalition by intervening.

Recall that while the third party does not know the partner's costs for staying in the coalition, it does know the distribution from which P's costs are drawn and the values of the cutpoints, $\underline{c}_P$ and $\hat{c}_P$, that partition the type space. As shown in Figure 5.2, this allows it to become increasingly optimistic that intervention will split the coalition as revealed preferences become more similar – that is, as the coalition grows more homogeneous. Since the partner's costs for remaining inside the coalition are uniformly distributed, or $c_P \sim U(0, \bar{c}_P)$, the probability that the coalition splits following the third party's intervention is

$$\Pr(c_P > \underline{c}_P) = \frac{\bar{c}_P - \underline{c}_P}{\bar{c}_P} = 1 - \frac{\beta \left[b_{LP}(1) - b_L(1) \right]}{\bar{c}_P}, \tag{5.7}$$

which turns out to have a direct role in the third party's alignment decision. While B is uncertain over the partner's willingness to stay in the coalition, it leverages its knowledge about the similarity of coalition members' revealed preferences, captured by β in Equation (5.7): the more positively correlated P's interests are with L's, all else being equal, the more durable the coalition in the absence of intervention *and*, as stated in Proposition 5.4, the more likely intervention will be effective at splitting it.

Proposition 5.4. *The probability that P is of a type that stays in the coalition is higher if B remains neutral than if B balances, such that* $\Pr(c_P \leq \underline{c}_P) < \Pr(c_P \leq \hat{c}_P)$, *as long as α is sufficiently small and δ sufficiently large.*

This effect of balancing on coalitional durability emerges only in equilibrium, as a consequence of intervention's effects on coalition members' remaining available resources. It also means that the third party's alignment choice has *three* effects on the coalition. First, it determines the probability that the coalition loses today's war, which eliminates it as a threat. Second, it diminishes the resources that the lead state, possibly in league with the partner, can use against the third party in the future. Third, it influences the partner's decision over whether to remain in the coalition. However, B takes this decision uncertain over the coalition's durability, weighing the potential gains of balancing – defeating or, failing that, weakening or splitting the coalition – against the costs of doing so, which will be wasted if P turns out to have been a type that would have left the coalition on its own. Remaining neutral, on the other hand, saves both the direct (c_P) and opportunity $(1 - \delta)$ costs of intervention, but leaves the coalition more likely to win today's war *and* less likely to disband.

As stated in Proposition 5.1, the third party balances against a coalition when the direct costs of doing so are not too large, or when

$$c_B < \overline{p}_{LP} \left(\beta \, [b_{LP}(0) - b_L(0)]^2 + b_L(0)\overline{c}_P \right)$$
$$- \underline{p}_{LP} \left(\beta \, [b_{LP}(1) - b_L(1)]^2 + b_L(1)\overline{c}_P \right) \equiv \hat{c}_B,$$

which reflects the same logic of war due to shifting power that animates Inequality (5.5) (cf. Powell 2004). However, in the coalitional case, the potential shift in power entailed by remaining neutral must also be conditioned by the probability that the coalition disbands; as shown by Equation (5.7), this is a function of both revealed preference similarity (β) and B's uncertainty over P's costs for cooperation (represented by $\overline{c}_P$). The added factor of coalitional durability also conditions the effect of the coalition's military power, in particular the probability that it defeats the target if B remains neutral $(\overline{p}_{LP})$.

Coalitions likely to be durable create for the third party an incentive to facilitate their defeat by joining the war against them in the short term, but as shown by Figure 5.2, they are also more likely to fall apart after winning today's war if B intervenes against them. However, as stated in Proposition 5.5, the effect of preference similarity on the intervention constraint also depends on the coalition's chances of victory, especially the extent to which intervention changes it.

Proposition 5.5. *B's intervention constraint increases in β, or $\partial \hat{c}_B / \partial \beta > 0$, when $\overline{p} > \overline{p}_{LP}^{\dagger}$, and it weakly decreases in β, or $\partial \hat{c}_B / \partial \beta \leq 0$, when $\overline{p} \leq \overline{p}_{LP}^{\dagger}$.*

The effect of the coalition's preference similarity on B's willingness to balance depends on its effect on the coalition's military prospects. Regardless of its choice, B becomes worse off as the coalition's preferences grow more similar, because preference similarity makes the coalition more likely to remain intact for future bargaining. Formally, the third party's expected utility decreases in β, but the rate of decrease depends on how intervention affects the outcome of the war. When the coalition is sufficiently likely to win today's war, or $\overline{p}_{LP} > \overline{p}_{LP}^{\dagger}$, where

$$\overline{p}_{LP}^{\dagger} = \underline{p}_{LP} \frac{(b_{LP}(1) - b_L(1))^2}{(b_{LP}(0) - b_L(0))^2},$$

then intervention has a large impact on the outcome, since the difference $\overline{p}_{LP} - \underline{p}_{LP}$ is relatively high. Covering the costs of intervention is easy, and B's payoff for intervention falls more slowly than its payoff for neutrality as the coalition becomes more durable. Thus, the intervention constraint becomes easier to satisfy as β increases. On the other hand, when the coalition is less likely to win today's war, or $\overline{p}_{LP} \leq \overline{p}_{LP}^{\dagger}$, intervention has a smaller impact on the outcome, making the costs of intervention difficult to cover as the coalition becomes more durable and less likely to disband. Thus, the payoff for intervention decreases more quickly in preference similarity than does the payoff for neutrality, and B becomes more likely to remain neutral as preference similarity increases.

To see why, consider first a coalition relatively unlikely to win today's war, or $\overline{p}_{LP} \leq \overline{p}_{LP}^{\dagger}$. Lower values of $\overline{p}_{LP}$ mean a smaller difference $\overline{p}_{LP} - \underline{p}_{LP}$, so intervention has a limited effect on the outcome of today's war. In other words, the coalition is already unlikely to win, so intervention cannot change the likely outcome very much. This minimal effect on the outcome makes the direct costs of intervention difficult to cover, a difficulty made worse as β increases, or as the coalition becomes more durable and poses a larger threat. As a result, intervention is more attractive when coalitions are relatively diverse in their preferences, because the costs are more easily surmounted; the coalition is unlikely to stay together in the unlikely event that it wins, and the third party is willing to pay the direct costs of intervention in order to drain resources from its future opponents. However, when coalitional preferences are more similar, then the coalition is more likely to remain intact even if B intervenes, and the third party foregoes a minimal effect on the outcome of the war in order to save the direct costs of intervention *and* safeguard its future bargaining power, avoiding the opportunity costs of intervention $(1 - \delta)$ against what will likely be an intact coalition. Thus, when the coalition is weak, the third party remains neutral against homogeneous coalitions but intervenes against diverse ones.

Now consider a coalition relatively likely to win today's war, or $\overline{p}_{LP} > \overline{p}_{LP}^{\dagger}$. Since $\overline{p}_{LP} - \underline{p}_{LP}$ is large, intervention has a greater impact on the probability that the coalition loses. Here, increasing preference similarity within the coalition encourages the third party to balance in hopes of defeating or splitting the coalition, and the intervention constraint becomes easier to meet as β increases. Put differently, as β decreases and the coalition becomes increasingly likely to disband on its own, the third party is more inclined to save the costs of intervention, since paying those costs would be wasteful – especially if it would like to husband resources, avoiding the opportunity costs of fighting, to face a powerful enemy in the future. Should a third party see a powerful, homogeneous coalition, it knows that the coalition is more likely to remain intact than a diverse one, but it also knows that it is more likely to be split by intervention than a diverse coalition. Thus, intervening against a powerful coalition – say, one led by a great power – is most attractive when its members have similar preferences. When coalitions are likely to defeat their targets, third parties intervene when coalitional preferences are relatively similar and remain neutral when coalitional preferences are more diverse.

Therefore, the model shows that there is an interactive, conditional effect between the similarity of revealed preferences and the coalition's military strength: when the coalition is sufficiently likely to win today's war, an increasing similarity of preferences should increase the probability provoking intervention, but it should diminish the probability of intervention when the coalition is weaker relative to its target. In other words, the threatening nature of coalitions should be tied directly to a third party's expectations over the coalition's durability, which affects incentives to balance. If partner choice plays such a significant role in both the durability of coalitions and their propensity to provoke balancing, then it seems reasonable that lead states consider these downstream effects when choosing between unilateral and coalitional action, and the following section explores that potential relationship in detail.

Coalition Formation

Moving back up the game tree in Figure 5.1, we can now consider the lead state's initial choice over acting unilaterally and taking on a coalition partner, whose attractiveness must be weighed against the probability that it provokes balancing and that it will continue cooperating with the lead state after today's conflict. In addition to generating further insights into which partners lead states accept and which ones they do not (and why), this section also clarifies some critical parts of the data generating process, particularly how the anticipation of balancing and coalitional durability affect choices over coalition-building – that is, the selection process by which coalitions emerge in the samples used for this chapter's empirical models. I show that, in equilibrium, the effects of preference similarity and military power both have non-monotonic effects on the attractiveness of acting multilaterally. However, increases in the probability that cooperating with P will provoke balancing, whatever parameters conspire to raise that probability, straightforwardly decrease the probability of coalition formation.

When considering whether to take on a particular partner, the lead state weighs the costs of ensuring P's cooperation against the likely effects of coalition-building on the probabilities of victory, balancing, and coalitional durability. This calculation is similar to those described in Chapters 3 and 4, where coalitions form in the event that their net benefits outweigh the costs; however, in this case the costs come not only in the form of direct compensation or adjusted mobilization decisions but also the entry of a third-party state into the war (Gartner and Siverson 1996, Werner 2000).

If the lead state acts alone, it saves the costs of securing P's military cooperation, though by refusing assistance it tolerates a greater risk of defeat in today's conflict, a risk made all the larger if the third party chooses to balance against it. If, on the other hand, L builds a coalition with P, it brings more military capabilities to bear on its side, but doing so may provoke balancing that it otherwise would not face had it acted as a singleton. As stated in Proposition 5.1, L builds a coalition when the costs of purchasing a partner's cooperation are not too large, or when $s < s^\dagger$ (characterized fully in the appendix), which turns out to be a complex function of L's beliefs over the third party's costs for intervention (captured by $\bar{c}_B$), preference similarity with its potential partner, and the distribution of military power. However, none of these critical parameters exercises a consistent, monotonic effect on the attractiveness of coalition-building.

Unsurprisingly, given the complex relationships between parameters that shape B's alignment choice, the decision to take on a coalition partner in the shadow of questions over balancing and coalitional survival is a complicated one. Depending on the values of other variables in the model, *any* of the parameters listed above, from the highest costs for intervention that B might pay to the similarity of preferences between L and P to measures of relative military power, can raise *or* lower $s^\dagger$ and make coalition formation with a particular partner either more or less likely. In part, this is because at many given values of any one parameter, arbitrary values of the others can be part of a set of conditions sufficient to produce balancing. This prevents the drawing of clear implications for the effect of any one variable across the stages of coalition formation and balancing. As stated in Proposition 5.6, however, the threat of balancing itself, however it emerges, *does* discourage L from forming coalitions with particular partners.

Proposition 5.6. *All else being equal, L's willingness to take on a particular partner decreases in the probability that P provokes balancing, since $s^\dagger$ decreases in $\hat{c}_B$.*

By treating the probability of balancing in reduced form, or $\hat{c}_B/\bar{c}_B$, we can gain a sense of its effects on L's willingness to build a coalition, whatever combination of the parameters might raise or lower the probability. Proposition 5.6 states that an increase in $\hat{c}_B$, or the type of third party that is just indifferent between balancing and remaining neutral, makes the lead state's coalition-building constraint ($s^\dagger$) harder to satisfy. On the other hand, coalitions are more likely to form with a potential partner when the probability with which that particular partner provokes balancing is low. More generally, L will find particular partners attractive only when the costs of securing their cooperation, often considerable on their own, are not compounded by the likely provocation of third-party opposition. Otherwise, the lead state balks at forming coalitions with states that are sure to draw in additional states on its target's side.

Two examples from coalitional wars in the latter half of the twentieth century are instructive here. First, in 1950, as it built a military coalition under UN auspices to overturn the North Korean invasion of South Korea, the United States turned down earnest offers of military assistance from the Nationalist government on Taiwan. Concerned that a Nationalist presence in a victorious coalition would make promises not to expand the war to China incredible (Stueck 1995, pp. 56–57) – an avowed goal of Chiang Kai-Shek's Republic of China government after its retreat from the mainland in 1949 – the United States not only turned down military assistance but also neutralized the Taiwan Strait, hoping to ensure that the Communist government of mainland China would be less inclined to intervene in the Korean War out of fear of a renewal of the Chinese Civil War (Stueck 2004, p. 223).[12] Likewise, in the 1991 Gulf War, the United States also ensured that Israel would remain neutral (Atkinson 1993, pp. 92–93), easing fears in Arab states, fears that might have led to balancing against the coalition, that a war against Iraq would not be used as a springboard for an expansion of Israeli interests in the region.

This also implies that when more homogeneous alternatives would otherwise provoke balancing, potential lead states may aim to build coalitions that are not durable – at least in the sense of maintaining cooperation that would pose threats to third-party states after winning today's war. Absent a threat of provoking balancing, however, durable coalitions that would merely maintain peace between their members after the war would be quite attractive. Therefore, it is difficult to draw strong empirical implications over the initial selection of coalition partners and their subsequent durability; however, the general approach to measuring durability that I employ in the following section should be relevant to the maintenance of cooperation in general, not just with respect to specific third parties after today's conflict. Nonetheless, potential selection bias is unlikely to be of substantial concern at the stage of coalitional durability.

Since lead states avoid coalition partners that are the most likely to provoke balancing, we should be careful in the interpretation of the empirical models of balancing in the following section, which are based on a set of observed coalitions. Specifically, Proposition 5.6 suggests that those coalitions most likely to provoke opposition and expand into multilateral conflicts will also be those least likely to form. Rather, lead states will either limit the size of the coalitions they do build or act as singletons as a way to keep the war limited in size. As discussed later, this should bias the empirical models that follow against finding strong relationships between their explanatory variables and a higher risk of balancing – which increases confidence in those relationships that are uncovered despite the likely attenuation of their effects.

Summary of Empirical Implications

Fearful third parties face a dilemma when choosing whether to pay the costs of war and join ongoing conflicts in order to balance against potential future threats. They are more inclined to balance against threatening singletons that

---

[12] The PRC would eventually intervene in the war, though not because it feared a joint American-ROC attempt to conquer it in the future. On the American desire to restrain the ROC more generally from attempting to retake the mainland, see Benson (2012, ch. 7).

are likely to defeat their targets in the present than those likely to suffer defeat at the hands of their targets. However, when judging the attractiveness of balancing against coalitions of states, third parties must also be mindful of the probability that the threat dissipates through the disbanding of the coalition. While coalitions with similar preferences are likely to maintain military cooperation in the future, those with more diverse preferences are more inclined to disband, in which case the threat dissipates. In equilibrium, a third party uses a combination of coalitional power and preferences to gauge the value of intervention; while an increasing diversity of preferences discourages intervention against coalitions likely to defeat their targets, increasing diversity actually encourages intervention against coalitions with poorer prospects of victory in today's war. Finally, any relationship between power, diversity, and balancing in coalitions is likely to be attenuated in observational data, because lead states will – if they can – avoid those partners that are most likely to undermine the value of military multilateralism by provoking a balancing response from third parties. The next section presents an empirical evaluation of these implications.

5.3 EMPIRICAL MODELS OF EXPANSION AND DURABILITY

When do coalitions provoke outsiders to intervene on behalf of their targets, and how do these patterns differ from singletons? The desire to balance – that is, intervening today in order to affect the future distribution of power (Powell 1999) – is one plausible explanation, especially in the context of coalitions and their aggregated military power. However, while Chapter 2 showed that coalitions are more likely to provoke intervention on their target's behalf than are states acting alone in a simple bivariate analysis, the associated decomposition analysis also showed that coalitions' uniquely high levels of aggregate power can account only for a small share of that difference. The theory in this chapter suggests that a crucial missing element in that difference is the diversity of preferences within coalitions, which affects their durability after victory. As such, there should be an interactive effect between the coalition's relative power and the distribution of preferences within it. In this section, I first leverage the data described in Chapter 2 to create a crisis-level dataset of conflict expansion, focused on when third parties join a target's side during a conflict, from 1946–2001. Then, since coalitional durability should be linked to both preference diversity and the occurrence of balancing, I analyze the duration of cooperation between the members of war-winning coalitions, using Correlates of War data on war participants and outcomes (Sarkees and Wayman 2010).

5.3.1 Hypotheses

The theoretical model implies relationships over two outcome variables: (a) third-party decisions to join the target's side in a conflict and (b) the durability ʋooperation between coalition members if they win today's conflict. Before outlining the research design appropriate to each, I translate five empirical implications derived from the theory into the language of their appropriate empirical models. One key implication – the avoidance of some partners that would surely provoke balancing – is important chiefly for the implied strength of the relationships posited for coalitions in the following hypotheses. As such, I do not list it separately but take care to use it to interpret those results more directly amenable to testing in this chapter's empirical models.

First, I test three hypotheses over the links between coalition-building, power, and preferences on the one hand and the occurrence of conflict expansion on the other, following the logic of Propositions 5.2 and 5.5. Consistent with the preventive war motivations of balancing, I first expect that singletons will be more likely to provoke balancing when they are more likely to defeat today's target.

Hypothesis 5.1. *When side 1 is a singleton, the probability that the conflict expands increases in side 1's chances of defeating its target.*

Power, however, should work differently when it is aggregated, because its persistence is also a function of the diversity of coalitional preferences. Relatively powerful, diverse coalitions should be less likely to provoke balancing than powerful but less diverse coalitions, while relatively weak and diverse coalitions should be more likely to provoke balancing than their relatively weak but homogeneous counterparts.

Hypothesis 5.2. *When side 1 is a coalition and sufficiently unlikely to defeat its target, the probability that the conflict expands increases in the coalition's preference diversity.*

Hypothesis 5.3. *When side 1 is a coalition and sufficiently likely to defeat its target, the probability that the conflict expands decreases in the coalition's preference diversity.*

Recalling the initial comparison of rates of conflict expansion explored in Chapter 2, Hypotheses 5.1–5.3 also bear on the conditions under which coalitions should be more and less likely to provoke balancing than singletons should. Suppose, first, that we can observe a singleton and a coalition at equal levels of relative military power – that is, equally likely to defeat the target in war. Then, note that if we consider a singleton to be a coalition of one, it must also have the least diverse preferences possible. As such, the probability that the coalition provokes balancing should be statistically indistinguishable from a fully homogeneous coalition; however, as coalitional diversity increases, the effects predicted by Hypotheses 5.2 and 5.3, as well as differences in the rates of conflict expansion across coalitions and singletons, should begin to emerge. Evidence of these conditional differences would provide further support for the claim that coalitions are not merely the sum of their aggregated parts.

Hypothesis 5.4. *When relative power is held constant, coalitions sufficiently unlikely to defeat their targets become more likely than singletons to see conflict expansion as diversity increases. Meanwhile, coalitions sufficiently likely to defeat their targets become less likely than singletons to see conflict expansion as diversity increases.*

As discussed earlier, the selection process identified in the theoretical model suggests that lead states will do their best to avoid coalition partners that provoke balancing. Therefore, the sample of crises analyzed later in the chapter will be biased against finding any strong relationship between power, preferences, and balancing. However, to the extent that these relationships are still recovered in observational data, it allows us to be more confident in the strength of the relationship, since the empirical test will be conservative.

Finally, Propositions 5.3 and 5.4 make predictions over the durability of the coalition should it emerge victorious from today's conflict. In the sections that follow, I adopt a very inclusive definition of cooperation in order to operationalize durability – specifically, the length of time that victorious war coalitions can survive without any of their members falling into open warfare with one another. As such, the following hypotheses are stated in terms of the duration of cooperation.

Hypothesis 5.5. *The length of time that a coalition maintains cooperation after victory will decrease in the diversity of preferences among its members.*

Hypothesis 5.6. *The length of time that a coalition maintains cooperation after victory will be shorter if it provoked balancing than if it kept its target isolated.*

Like Chapter 3, I test the empirical implications of this chapter's theoretical model against two distinct dependent variables. However, given my interest in both what happens during conflicts and what happens afterwards, as well as some data limitations with respect to the latter, I present the two research designs in separate sections, each one built around a distinct set of data and a unique set of estimation challenges. I test Hypotheses 5.1–5.4 on conflict expansion first, then turn to Hypotheses 5.5 and 5.6 before discussing the cases of Afghanistan and Iraq that opened the chapter.

5.3.2 Research Design: Conflict Expansion

To create the appropriate dataset for testing Hypotheses 5.1–5.4, I begin with the list of ICB crises (Wilkenfeld and Brecher 2010) for which I coded coalition formation in the period 1946–2001 (see Chapter 2). Next, I aggregate participants by side, 1 or 2, in order to create the coalition-level variables implied by the theory: the similarity/diversity of revealed foreign policy preferences, side 1's chances of defeating T militarily, and the occurrence of third-party military support for the target. This results in a set of directed side-dyads, such that each side has the opportunity to be measured as both a potential coalition (side 1) and target (side 2), as was the case in Chapters 3 and 4. There are thus 460 directed crisis observations, 67 of which involve a coalition on side 1 and 393 in which side 1 is a singleton.

Recall that my operational definition of coalition excludes states that join side 1 after the crisis escalates to war. However, for this analysis focused on conflict expansion, third-party intervention on the target's side can take the form of a counter-coalition forming against side 1 before war breaks out *or* states joining side 2 (but only side 2) after war begins. Thus, the variable Balancing equals zero if the target faces side 1 alone and one if the target receives military cooperation, as defined in Chapter 2, from at least one third party, before or after the crisis escalates to war.[13] It is important to note that the structure of the data assumes that *some* third party is tempted to balance in every crisis. Thus, it is possible that the data include observations that are not "relevant" in the sense that they do not satisfy the assumption that some third party fears future interactions with crisis participants. To the extent that this is true, the inclusion of these cases only introduces noise into the analysis, rendering hypothesis tests on the key theoretical variables more conservative. If significant results are found despite this noise, then it should increase our confidence in the results. Thus, I leave them in the data.[14]

The key theoretical variables implied by the theory are side 1's probability of defeating the target and the diversity/similarity of side 1's revealed preferences. To measure side 1's probability of defeating the target, I follow the theoretical model and construct a measure of side 1's probability of defeating its primary target state on side 2 (identified in Chapter 2), since third parties make intervention decisions based on the target's probability of victory absent any assistance. The resultant variables, $\Pr(\text{Win}_{LP})$ for coalitions and $\Pr(\text{Win}_L)$ for singletons, use military capabilities as measured by the Correlates of War project's CINC index (Singer, Bremer, and Stuckey 1972) to construct the following probability of victory for side 1,

$$\Pr(\text{Win}) \equiv \frac{m_L + \sum_0^n m_P}{m_L + \sum_0^n m_P + m_T} \equiv \frac{\text{CINC}_1}{\text{CINC}_1 + \text{CINC}_T}.$$

Following the theoretical models of this and the preceding chapters, side 1's probability of defeating its primary target militarily is side 1's share of the total military capabilities in the dyad. Thus, side 1's probability of victory increases

---

[13] While it is possible to code the actual number of observers that join the target, the model makes predictions only over the existence of outside support from one third party. We can imagine that the presence of some support may deter others from joining, which would imply inconsistent effects for the independent variables across different values of the dependent variable. Therefore, I restrict the analysis only to the presence of third-party support for the target, not to the number of states offering such support.

[14] Further, since I measure no variables on potential balancers, the problem of irrelevant observations is substantially less than it otherwise might be.

in its relative share of the total military capabilities in the crisis and decreases in the target's share of capabilities (cf. Powell 1999, ch. 5). Notably, this measure excludes the capabilities of any members of side 2 beyond the initial target, which would be highly correlated with the outcome variable and potentially overstate the model's ability to predict balancing.

Next, Diversity$_1$ is the inverse of β in the theoretical model; it increases in the heterogeneity rather than similarity of revealed foreign policy preferences within a coalition. As Diversity$_1$ increases, the coalition's revealed foreign policy preferences grow more divergent (i.e., low β), and its preferences grow more similar as Diversity$_1$ decreases (high β). As before, I construct this measure with UNGA ideal point data (Strezhnev and Voeten 2013).[15] The closer are two countries' ideal points, the more similar are their revealed preferences and the more likely they are to agree on, and cooperate over, a given issue in the future. After identifying each coalition member's ideal point in the year prior to the crisis to avoid problems of reverse causality, I then calculate the variance in those ideal points to proxy for the diversity of preferences within a coalition. Since I split the sample into coalitions and singletons for side 1, this variable is measured only on coalitions, resulting in a variable that ranges from 0 to roughly 3.29, although the mean for coalitions is rather low, about 0.09, indicating that there are few coalitions with highly diverse preferences – as we should expect if compensating partners with divergent preferences is expensive (see Chapter 3).[16]

Since Hypotheses 5.1–5.4 entail comparisons across singletons and coalitions, an empirical model estimated on the full sample would entail a triple interaction term involving (a) the presence/absence of a coalition, (b) side 1's probability of defeating T, and (c) the diversity of preferences within coalitions. Assuming that allocating to singletons a value of 0 for preference diversity is valid – and, given the results of the decomposition analysis of Chapter 2, it probably is not – this triple interaction involving two continuous variables, as well as every possible combination of components, would create enough multicollinearity to place a strain on the small size of the sample. Thus, I opt instead to split the sample into singleton and coalition observations, which facilitates the interpretation of the interaction terms and eases comparison of the relevant theoretical variables across both subsamples.

Several additional factors related to coalition formation, power, and diversity might confound the predicted relationships, so I account for a number of

---

[15] I present results using Strezhnev and Voeten's (2013) data; results are generally consistent using Reed et al.'s (2008) ideal points, which are also used in Wolford (2014a).

[16] I have also conducted the analysis with an alternative measure that weights ideal points by military capabilities, but it produced no substantive change in the results. However, to capture the relative importance of "small" states in coalitions, such as Saudi Arabia in the Gulf War or Kuwait in the Iraq War, I use the unweighted measure in the main analysis. For a similar analysis conducted on a sample of both coalitions and singletons, see Wolford (2014a), which also includes a larger set of control variables.

them as control variables. First, since side 1's probability of military victory is a reduced form representation of each side's military capabilities, I include both side 1's total military capabilities, $CINC_1$, as well as those of the target state, $CINC_T$ (Singer, Bremer, and Stuckey 1972, Singer 1987). Both factors shape incentives to form coalitions and to balance, and including each allows for a consideration of the effect of relative power independent of the actual military size of the disputants. Next, $Number_1$ is a count of the states on side 1 when it is a coalition, which may be correlated with both preference diversity and/or power on side 1, as well as the provocation of opposition. I also include the square of the count of coalition members, $Number_1^2$, to account for the possibility that larger coalitions disband after victory independently of any related diversity of preferences, as posited by Riker's (1962) size principle; increasing size from low values may be associated with greater chances of victory, making balancing attractive, but once the coalition grows large enough, the fact of its likely dissolution may render its size associated with a reduction in incentives to balance. Finally, $Percent\ Allied_1$ is the percent of coalition member pairs that share an ATOP alliance tie (Leeds, Long, and Mitchell 2000), which makes coalition formation attractive and might presage particular patterns of postwar cooperation. These latter three variables are only measured on coalitions.

Several features of the conflict and the historical era may also be relevant. War equals 1 if the crisis escalated to "full-scale war" as identified by the ICB data (Wilkenfeld and Brecher 2010), and it is designed to capture the extent to which longer or more severe conflicts may involve a larger number of states. This is especially relevant to the extent that more powerful or diverse coalitions tend to fight longer or more severe conflicts.[17] To account for the proposed link between the approval of the United Nations Security Council on coalition formation (Chapman 2011, ch. 5) and discouraged balancing (Voeten 2005), $UNSC\ Support_1$ equals one if that body explicitly approves of side 1's use of force or condemn's T's actions in the crisis (see Chapters 3 and 4). However, its basic rarity and dependencies with several other variables in the singleton subsample result in this variable's omission by the statistical software, so its coefficient is estimated only for the coalition subsample.[18]

Next, $Land\ Borders_T$ is the number of states with which the target state shares land borders according to the COW contiguity data (Stinnett et al. 2002), while $Log\ Allies_T$ is the natural log of the number of states with which T shares a defense pact, according to the ATOP data (Leeds et al. 2002). Both variables may be related to other states' perception of a shared threat from the coalition as well as the availability of potential balancers; the more neighbors and the more allies a target state has, the more potential third parties there may be to come to its aid, and the more a lead state's decision over choosing partners

---

[17] The war indicator also accounts for the collapse of intra-war crises into single observations.

[18] All empirical models are estimated in Stata 13.1.

may be affected. Next, Min Democracy$_1$ is the Polity-IV combined democracy score of the least democratic state on side 1 (Marshall and Jaggers 2009), which accounts for the possibility that democratic states may be uniquely inclined to form coalitions and/or be adept at signaling limited aims to third parties considering intervention. Democracy$_T$ is the corresponding democracy score for the target state, which may be related to the target's ability to attract partners and, as such, the desire to form coalitions against it. Post Cold War equals 1 in all years after 1990 and 0 otherwise, to account for (a) the potentially greater threat posed by great power coalitions in a unipolar world and (b) the possibly greater effects of the Security Council on coalition-building and reassurance after the end of the American-Soviet superpower rivalry (Voeten 2005). Finally, United States$_L$ indicates whether the United States is the lead state on side 1, reflecting its position as a dominant military power, giving it unique potential to threaten observers, and its prolific use of coalitions.

Thus, the full statistical specification is

$$\Pr(\text{Balancing} = 1) = \Phi(\alpha + \beta_1 \Pr(\text{Win}_{LP}) + \beta_2 \text{Diversity}_1 + \tag{5.8}$$

$$\beta_3(\Pr(\text{Win}_{LP}) \times \text{Diversity}_1) + \beta_k \mathbf{X}_{si} + \varepsilon_i),$$

for coalitions and

$$\Pr(\text{Balancing} = 1) = \Phi(\alpha + \beta_1 \Pr(\text{Win}_L) + \beta_k \mathbf{X}_{si} + \varepsilon_i), \tag{5.9}$$

for singletons, where Φ is the CDF of the standard normal distribution, implying a probit model, and $\mathbf{X}_{si}$ is a vector of control variables measured by side s and crisis i. Finally, given the "double-counting" of crises due to the directed-dyad form of the data, I use Huber-White robust standard errors to account for intergroup correlations and cluster them by the individual crisis.

5.3.3 Results: Conflict Expansion

Table 5.1 summarizes the results of estimating the probit models in Equations (5.8) and (5.9), where Model 1 estimates the probability of balancing on the coalitions subsample and Model 2 does the same for the singletons subsample. I begin with Hypothesis 5.1, which makes a prediction about balancing against singletons, then follow by exploring the interactive effects predicted for coalitions in Hypotheses 5.2 and 5.3, before closing with a consideration of Hypothesis 5.4, which compares the relative probabilities of balancing across the two subsamples.

Beginning with Hypothesis 5.1, we should expect the coefficient on a singleton's probability of victory, $\Pr(\text{Win}_L)$ in Model 2, to be positive, and this is indeed the case. The more likely singletons are to defeat their primary target, the more likely they are to provoke balancing, and this positive relationship is statistically significant; it is distinguishable from a null relationship at $p < 0.01$. This lends some support to the notion that, when they join ongoing conflicts,

TABLE 5.1. *Probit Models of Conflict Expansion for Coalitions and Singletons, 1946–2001*

	Pr(Balancing $= 1$)	
Variable	Model 1 Coalition$_1$	Model 2 Singleton$_1$
– Theoretical variables –		
Pr(Win$_L$)	–	1.09 (0.39)***
Pr(Win$_{LP}$)	−2.55 (0.91)***	–
Diversity$_1$	15.03 (3.53)***	–
Pr(Win$_{LP}$)× Diversity$_1$	−16.31 (3.84)***	–
– Control variables –		
CINC$_1$	5.67 (5.07)	−0.91 (4.29)
CINC$_T$	26.77 (8.23)***	9.59 (2.48)***
Number$_1$	6.66 (2.11)***	–
Number$_1^2$	−0.79 (0.28)***	–
Percent Allied$_1$	1.35 (0.95)	–
War	4.94 (1.16)***	1.11 (0.28)***
UNSC Support$_1$	1.31 (0.84)	–
Land Borders$_T$	−0.76 (0.18)***	−0.02 (0.04)
Log Allies$_T$	0.01 (0.08)	0.01 (0.03)
Min Democracy$_1$	−0.01 (0.05)	−0.02 (0.01)*
Democracy$_T$	−4.28 (0.99)***	−0.19 (0.21)
Post Cold War	−2.09 (0.88)**	0.18 (0.22)
United States$_L$	3.54 (1.31)***	−0.47 (0.73)
Intercept	−10.45 (3.05)***	−2.02 (0.33)***
Model Statistics		
N	67	393
Log-likelihood	−16.87	−128.06
$\chi^2_{(d.f.)}$	55.05$_{(16)}$***	37.01$_{(10)}$***

Significance levels :   * : 10%   ** : 5%   *** : 1%

third parties do so in hopes of shifting the distribution of power against states that they view as potential future threats. However, Hypotheses 5.2 and 5.3 predict that the relationship between relative power and balancing works rather differently when side 1 is a coalition.

Moving to Model 1, estimated on the coalitions subsample, recall that Hypotheses 5.2 and 5.3 expect that the probability of balancing will increase in Diversity$_1$ when side 1 is relatively powerful, but it should decrease in Diversity$_1$ when side 1 is relatively weak. However, these predicted relationships cannot be inferred directly from the coefficients estimated in Table 5.1 (Ai and Norton

2003, Brambor, Clark, and Golder 2006), a task made all the more difficult since both components of the interaction are continuous variables. The procedures used in Chapter 4 are designed to assess the effect a joint increase in the two variables (Norton, Wang, and Ai 2004), but this is less useful for Hypotheses 5.2 and 5.3, which make predictions over the effect of an increase in one variable, Diversity$_1$, at specific values of another, Pr(Win$_{LP}$). As such, I generate predicted probabilities of balancing for two types of coalition – relatively powerful and relatively weak – and plot them against changes in coalitional diversity in Figure 5.3.

The theoretical model predicts that an increasing diversity of coalitional preferences will decrease the probability of balancing for powerful coalitions and increase it for relatively weak coalitions. The simulations in Figure 5.3 compare the effect of increasing the diversity of side 1's revealed preferences on coalitions that are "average" in every respect but their aggregate military power. According to the data, the representative coalition consists of two members, has a maximum democracy score befitting a mixed regime (neither fully democratic nor fully autocratic), does not involve the United States, and is ignored by the UN Security Council.[19] The simulated coalition on the left has a probability of defeating its target that is one standard deviation below the sample mean, while the coalition on the right has aggregate military power one standard deviation above the mean, representing (roughly) a coalition of two minor powers at left and a coalition involving at least one great power at right. For each simulation, Diversity$_1$ ranges from complete homogeneity to the sample maximum.

The black line in the left panel of Figure 5.3 is the predicted probability that side 1 provokes balancing, with dashed 95% confidence intervals to either side. As anticipated by Hypothesis 5.2, this probability begins low and increases – quite rapidly – through increasing levels of preference diversity. Despite the right-skewed distribution of Diversity$_1$, there is a statistically and substantively significant increase in the probability of provoking balancing, from roughly nil for homogeneous coalitions to near certainty for highly diverse coalitions. Since third parties have only a small impact on the chances of victory for weak coalitions, they tend to remain neutral against homogeneous coalitions, saving their own capabilities rather than waste them on an outcome unlikely to change; intervening against diverse coalitions is more attractive, however, because they can be more easily split as the result of intervention.

Moving to the more powerful coalition on the right, there is a less substantial though statistically discernible decrease in the probability of balancing as the coalition's preferences grow increasingly diverse. Here, third parties have a greater impact on the probability of victory than when they face weak coalitions, leading them to prefer intervention against homogeneous coalitions, which are less likely to disband in the event of victory. The 95% confidence

---

[19] All continuous control variables are held at their means, and all dichotomous control variables at their modes.

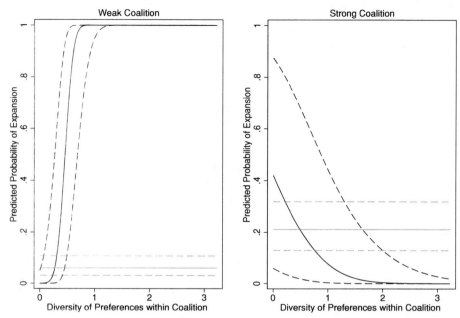

FIGURE 5.3. Predicted probability of balancing for coalitions and singletons by probability of military victory and diversity of preferences with 95% confidence intervals

interval narrows at empirically rare high levels of diversity, but the overall statistical strength of the relationship is difficult to assess, given the rarity of coalitions that are both very diverse and very powerful. Moving from a minimally to a maximally diverse coalition results in a large decrease in the probability of provoking balancing, from Pr(Balancing = 1) = 0.45 to nearly Pr(Balancing = 1) = 0. On the other hand, if power alone affected the incidence of opposition – that is, if observers opposed any and all concentrations of power – then the slope of the line at right in Figure 5.3 should be both flat and everywhere higher than a similarly flat line at left. However, the predicted probabilities each display nonzero slopes, each moving in different directions, as would be the case if there were an interactive effect between power and diversity. Therefore, consistent with the intuition of the decomposition analysis in Chapter 2, their aggregate military power alone cannot account for coalitions' unique propensity to provoke opposition; the interaction of power and preferences, on the other hand, can go some way to explaining the conditions under which balancing does and does not occur.

What do diverse and homogeneous coalitions look like in the empirical record? Two of the least diverse coalitions in the sample are the Soviet Union's coalition with the Afghan government following the 1978 coup (ICB #303) and the Saudi-Jordanian coalition in the Yemeni Civil War in 1962 (#195). Some of the most diverse are the Franco-Chadian coalition that sought to end Libyan support for insurgents in 1983 (#332) and the Kenya-Ethiopia coalition that resisted Somali attempts to capture the Ogaden region in 1980–1981 (#320). The theoretical model predicts that the latter two coalitions, both relatively powerful with a record of divergent preferences and thus more likely to disagree over future policy than the former, should be less likely to provoke balancing. Indeed, neither of the more diverse coalitions provoked third parties to support their targets, while the relatively powerful Saudi-Jordanian and Soviet-Afghan coalitions drew opposition in the form of outside support for the targets of their coercive efforts.

Next, Figure 5.3 is also consistent with the prediction of Hypothesis 5.4, which predicts that coalitions and singletons should experience similar probabilities of balancing when their power and diversity are equal, before diverging in ways predicted by Hypotheses 5.2 and 5.3 as conditions grow more diverse. As such, the gray lines in each panel plot the predicted probability of balancing, with 95% confidence intervals, for a singleton at the same level of relative power as the simulated coalition. In each case, all other relevant variables are also held at coalitional means and modes. Beginning in the left panel, the probability that equally powerful coalitions and singletons provoke balancing is roughly similar when coalitions are relatively homogeneous; however, as $Diversity_1$ increases, weak coalitions are more likely to provoke balancing than their equally powerful singleton counterparts. In the right panel, the story is similar; at low levels of $Diversity_1$, coalitions and singletons provoke balancing at a similar rate, but increasing diversity eventually helps powerful coalitions discourage balancing – through a credible promise to disband – and drop the predicted probability below that of equally powerful singletons.

Figure 5.3 also shows graphically that, as predicted by Hypothesis 5.1, the probability of balancing increases in a singleton's probability of victory; the probability of balancing is higher in the right panel, and since the 95% confidence intervals do not overlap, the difference is statistically discernible. This result, as well as the stark differences discovered as a function of changing levels of diversity, supports the contention that coalitions are not mere aggregations of power; if they were, the predicted probabilities of balancing for coalitions would simply remain flat. Just as they did in the empirical models of Chapter 4, coalitions respond to, or provoke different responses to, the distribution of power than singletons do. While, on average, coalitions provoke balancing at a greater rate than do singletons, that average effect masks a significant amount of variation across the two subsamples – variation that can only be uncovered by an explicit accounting of preference diversity inside coalitions.

Further, the control variables in Table 5.1 have implications for other models of coalitional politics. First, the percentage of allied states inside coalitions is unrelated to the probability of conflict expansion, perhaps challenging notions that allied coalitions invariably provoke counter-coalitions (see Vasquez 2009). Second, there appears to be a nonlinear relationship between the number of states on side 1 and the incidence of third parties balancing against them, rising through low numbers and falling through higher ones, that is consistent with Riker's (1962) size principle. Specifically, overlarge winning coalitions are more likely to disband than smaller ones are, which reassures potentially fearful observers that the coalition will not pose a long-term threat. Thus, large coalitions appear to provoke opposition less often than small coalitions do, though this does not affect the main results over power and diversity.

Next, recall that UNSC Support$_1$ indicates whether side 1 received support from the Security Council. Support from the UNSC may legitimize the use of force in the eyes of observer states, discouraging balancing by facilitating support or, at least, neutrality, reducing the pool of states available as potential balancers. The UNSC supports side 1 in roughly 11% of the crises in the sample, and its positive coefficient does not reach statistical significance. More importantly, however, the hypothesized interaction between power and diversity is robust to the inclusion of this control; even in the presence of what might be "easy" cases of global support for the coalition's aims, the theoretical model's predictions remain significant. Another possibility is that global consensus reflected in UNSC support may explain both the diversity of coalitions and any resulting negative effect on opposition. However, diversity *increases* the probability of opposition for weak coalitions, which is explicable in terms of the present theory and not by a relationship between diversity and global consensus. However, coalitions do appear to see a lower rate of balancing during the unipolar post Cold–War era, as indicated by the negative coefficient on Post Cold War, which is statistically discernible at $p < 0.05$, but its interaction with UNSC Support$_1$ fails to generate a significant relationship, making Voeten's (2005) claim about the unique effects of the UNSC after 1990 impossible to assess with these data.

In sum, the empirical model is encouragingly consistent with the expectations generated by the theoretical model. It also helps explain both the conditions under which coalitions provoke balancing and a key variable omitted from the decomposition analysis in Chapter 2: the diversity of foreign policy preferences within the coalition. To be sure, other coalition-specific variables might be relevant to explaining the difference between coalitions and single disputants, and this empirical model is in principle useful for identifying what those factors might be. Finally, the observed differences in rates of balancing as a function of relative military power across coalitions and singletons are consistent with the predictions of the theoretical model.

5.3.4 Research Design: Coalitional Durability

Hypotheses 5.5 and 5.6 make predictions not about the probability of balancing, but about the durability of coalitions after winning a war. As such, testing them requires a different statistical model, as well a different dataset, given the limited temporal scope of the coalitions data. In this section, I discuss the

construction of the relevant sample of victorious war coalitions, the collection
and coding of key theoretical and control variables, and the specification of
the Cox proportional-hazards model used to estimate the time between victory in war and the end of cooperation between former coalition partners.
The event-history framework is ideal for analyzing the durability of coalitions
because it models the probability that a coalition stays intact, or that it "survives," given that it has survived until the present time (see Box-Steffensmeier
and Jones 2004). When any of a coalition's members fall into war with one
another, the coalition "fails," or exits the sample – a crude measure of coalitional breakdown but, as discussed later, one that has some advantages. Event
history techniques have been widely used to study the duration of war (Bennett
and Stam 1996, Goemans 2000, Slantchev 2004) and the duration of peace
between former belligerents (Fortna 2003, Lo, Hashimoto, and Reiter 2008,
Werner 1999, Werner and Yuen 2005), but I apply them here to the problem
of postwar peace between former partners.

Assessing the durability of post-victory cooperation requires, first, identifying those coalitions that win interstate wars. To do so, I use the Correlates of
War project's list of interstate wars from 1816–2010 (Sarkees and Wayman
2010), which identifies twenty-five victorious coalitions in interstate wars, starting with France, Austria, and the Kingdom of the Two Sicilies' victory in the
War of the Roman Republic in 1849 and ending with the toppling of Iraq's
government during the interstate phase of the Iraq War in 2003 by the United
States, United Kingdom, and Australia.[20] Table 5.3, which can be found in the
appendix, lists the coalitions and their members that COW identifies as winning their wars, generated by eliminating all bilateral wars, as well as those in
which coalitions lose to single belligerents. The remaining wars are those won
by coalitions, as opposed to those in which coalitions lose or see the war end
in a stalemate. Some wars transition to another type of war, often intrastate
if conquest opens the door to insurgency; in these cases, if the interstate phase
had a clear winner – like the Iraq War, in which the coalition successfully toppled the target government before the emergence of the insurgency – I coded
the war's end at the transition point and recorded the outcome accordingly.[21]

Eleven coalitions, slightly more than half the sample after missing data is
taken into account, see a war break out among at least two of their members
after victory and before the end of the observation period, the first of which
for a coalition marks the end of cooperation and serves as the indicator of

---

[20] COW's operational definition of coalition membership differs from mine in some notable ways.
Kuwait, for example, from which the 2003 invasion of Iraq was launched, satisfies my definition
of a coalition member for the war, but not COW's.

[21] This leads to the inclusion of the War Over Angola (COW #186), Phase Two of the Second
Ogaden War (COW #187), the 2001 Invasion of Afghanistan (COW #225), and the 2003 Iraq
War (COW #227).

"failure" by which coalitions exit the data.[22] To be sure, coalitions may see cooperation break down in a number of ways short of full-scale war, but using war as a measure of coalitional dissolution has two advantages. First, all states can, in principle, go to war with one another, and they may do so over a variety of issues. Therefore, intramural war as a measure of coalitional breakdown does not depend on the issues involved in the war just ended or the specific terms imposed after victory; in contrast, indicators short of war – say, abrogated alliances (Leeds and Savun 2007, Leeds, Mattes, and Vogel 2009) – might. Second, as a relatively coarse indicator, war is also less prone to measurement error than some more inclusive alternative variables are, such as lower-level disputes or diplomatic tensions (e.g., Ghosn, Palmer, and Bremer 2004, Wilkenfeld and Brecher 2010); when war erupts between two states, there is little doubt that security cooperation between them has broken down.

There are two key theoretical variables: (a) the diversity of preferences within the coalition (Hypothesis 5.5) and (b) the existence of a balancing coalition on the other side in the war (Hypothesis 5.6). Measuring the diversity of preferences inside coalitions is difficult, as the limited temporal coverage of UNGA ideal points (Reed et al. 2008, Strezhnev and Voeten 2013), which are nonexistent before 1946, eliminates too much of the sample. Therefore, as a rough proxy of the diversity of preferences inside the coalition, I create the variable Diversity by calculating the variance of the Polity-IV combined democracy scores (Marshall and Jaggers 2009) among members in the first year for which data is available after the war, on the assumption that, in general, states with similar domestic institutions will hold similar preferences over foreign policy; the greater the variance in Polity scores, the more diverse the coalition's preferences. Given the large number of European wars in the data, linking Polity scores to what Braumoeller (2012) calls the distribution of ideology is far from inappropriate; in both the nineteenth and twentieth centuries, democracies and various forms of autocracy contested the relative sizes of the democratic and either legitimist or communist communities, making domestic political similarity a workable, if imperfect, indicator of preference similarity. Next, Balancing is an indicator variable that equals one if the victorious coalition defeated another coalition in the war, according to the COW data (Sarkees and Wayman 2010), and zero otherwise.

I also include four control variables. First, Number is a simple count of the states making up the victorious coalition, which may be linked to preference diversity, the occurrence of balancing (see Table 5.1), as well as the duration of cooperation, which affects durability through Riker's (1962) size principle: larger coalitions should also disband sooner than smaller ones. Second, Total CINC is the sum of CINC scores (Singer, Bremer, and Stuckey 1972) for each member of the coalition in the year after the war, since total power may be

---

[22] There are also no "ties," in that no wars erupt the same number of days after victory, making the choice of how to handle ties irrelevant.

related to both diversity (see Chapter 3) and the occurrence of balancing, as well as the sustainability of cooperation after the war; the more powerful a coalition, the more attractive it may be to keep together, and the more robust it may be to attempts to split it by draining its resources. Next, Percent Allied is the fraction of member pairs that share formal alliance commitments in the first year of the war (Leeds et al. 2002).[23] If, for example, a coalition includes three states, and two of the three possible pairs are allied in the year the war begins, then Percent Allied equals approximately 66.6; if all three states are allied with one another, then Percent Allied equals 100. Finally, Log Duration is the natural log of the length of the war, in days, as coded by Sarkees and Wayman (2010); long or severe wars may be the result of balancing (Shirkey 2012, Slantchev 2004), but they may also make future cooperation between their victors more difficult, to the extent that resources, capabilities, or the willingness to fight are drained by the length of the war.

Thus, the full specification of the Cox proportional-hazards model is

$$h(t|X) = h(t)\exp(\beta_1 \text{ Diversity} + \beta_2 \text{ Balancing} + \beta X_i), \qquad (5.10)$$

where the hazard function $h(t|X)$ characterizes the probability that a coalition survives intact at time t, given that it has survived up to time t, as a function of a set of regressors and their coefficients. The two theoretical variables are accompanied by a vector of control variables X_i, measured on individual coalitions i. The Cox model makes no assumptions about the underlying or baseline hazard rate, but it does assume that the regressors have a similar effect on the probability of survival at any time after victory – that is, that the hazards are proportional. Following Box-Steffensmeier and Jones (2004, pp. 124–137), I examine the Schoenfeld residuals as a function of time to look for possible violations of the proportional hazards assumption. I find none, which indicates that the model does not need to be adjusted for time-varying effects. I turn now to a discussion of the results of estimating this model on the sample of victorious war coalitions.

5.3.5 Results: Coalitional Durability

Table 5.2 presents the results of estimating the Cox proportional-hazards model from Equation (5.10), where I present hazard ratios rather than coefficients. The hazard ratio is the multiplicative effect on the hazard rate – that is, on the probability that a coalition dissolves at time t – of a one-unit increase in the variable, however units are scaled for that particular variable. If an increase in a given variable has no effect on the hazard, then the hazard ratio

---

[23] Using the Alliance Treaty and Obligations Project (ATOP) data (Leeds et al. 2002), I consider states allied if they share a defensive, offensive, or consultation pact. I also estimate the model using the Correlates of War data (Gibler and Sarkees 2004), where I code allies as those sharing a defense pact or an entente, and the results do not change.

TABLE 5.2. *Cox Proportional Hazard Model of Time Until War Between Victorious Coalition Partners, 1859–2007*

Pr(War/Failure = 1)	
Variable	**Estimates**
– Theoretical variables –	
Diversity	1.22 (0.11)**
Balancing	3.32 (6.03)
– Control variables –	
Number	8.86 (8.09)**
Total CINC	0.84 (0.07)**
Percent Allied	1.13 (0.07)**
Log Duration	0.48 (0.20)*
Model Statistics	
Subjects/failures	21/11
Log-likelihood	−13.73
$\chi^2_{(d.f.)}$	$22.27_{(6)}$***

Significance levels :   * : 10%   ** : 5%   *** : 1%

is 1; accordingly, an increase in the hazard rate (here, war sooner rather than later) implies a ratio greater than one, while decreases in the hazard rate (war later rather than sooner) are indicated by a ratio less than one. Further, while Table 5.3 identifies twenty-five victorious coalitions, data limitations restrict the sample to twenty-one coalitions, ending with Phase II of the Second Ogaden War (COW #187) in 1978, and the end of the observation period for all coalitions is 2007.

Beginning with Diversity, Hypothesis 5.5 anticipates that its hazard ratio should be greater than one, which would mean that more diverse coalitions tend to disband, falling into intramural conflict, sooner than less diverse coalitions. As indicated by the statistically discernible ($p < 0.05$) hazard ratio of 1.22, this is the case – even when the potentially confounding effects of coalitional size and strength are accounted for. The substantive effect is difficult to assess directly, given the scaling of the variable, which ranges from zero (the least diverse coalition) to eighty-five (the most diverse). A one-unit increase in the variance here makes little sense, so Figure 5.4 plots the estimated hazard function, which tracks the probability of coalitional breakdown over time as a function of a minimally diverse coalition, a coalition of average diversity, and a maximally diverse coalition. Beginning with the low-variance coalition, the probability of war between the victors begins low and increases only slightly over time, but moving to coalitions of average and sample-maximum diversity,

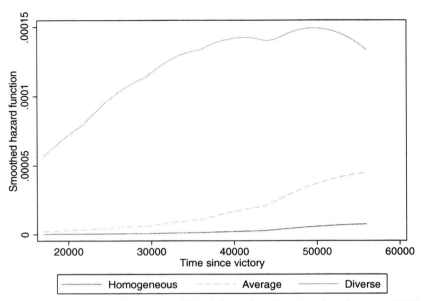

FIGURE 5.4. Hazard of coalitional breakdown by diversity of preferences, 1859–2007

each substantially increases the risk that victorious coalitions will break down into subsequent conflict.[24] Consistent with Hypothesis 5.5, homogeneous coalitions, regardless of size or aggregate strength, are uniquely durable compared to their more diverse counterparts. This may be good news for the victors down the line but, as shown earlier, potentially bad news for other states that worry about just how long the coalition might stay together and take advantage of its aggregate power.

Next, the hazard ratio on Balancing is of the expected magnitude, such that victorious coalitions that faced balancing appear over three times likelier than those who did not to break down at any given time. However, the coefficient fails to reach conventional standards of statistical significance; as such, the model and data do not produce support for Hypothesis 5.6. Nonetheless, there are reasons to be cautious about this result. First, as suggested by the theoretical model, lead states may only tolerate balancing when it is unlikely to threaten the coalition's durability, which would mask the true strength of the relationship in observational data. Second, the sample itself is small and the measure of coalitional breakdown quite crude; a finer-grained indicator of the fragility of

---

[24] Diversity ranges from 0 to roughly 85, with a mean of about 21 and a standard deviation of 29. As such, for the homogeneous coalition in Figure 5.4, Diversity = 10, Diversity = 21 for the average coalition, and Diversity = 40 for the diverse coalition (an increase of one standard deviation above the mean). All other variables are held at means or modes, as appropriate.

military cooperation may yet uncover a relationship that cannot be divined from the present empirical model.

Turning to the control variables, it is notable that larger coalitions see wars erupt between their members sooner than do smaller coalitions, in a pattern consistent with Riker's (1962) discussion of the size principle (see, in particular, chs. 2, 3). Thus, sharing the spoils of victory, regardless of any diversity in preferences, seems more difficult for larger coalitions than for smaller coalitions; with the losers defeated, large coalitions are valueless, because erstwhile victors have incentives to maximize their share of the remaining spoils by whittling down the former coalition to a new minimal winning one, which maximizes their share of the benefits – at the expense of some former partners.[25] Additionally, powerful coalitions appear more durable than weaker ones, and longer wars are followed by longer periods of peace between the victors than shorter wars are.

Most interesting among the control variables, however, is the relationship between alliances and the sustainability of post-victory cooperation. Specifically, an increase in Percent Allied is also associated with a statistically discernible ($p < 0.05$) increase in the risk of coalitional breakdown; in this case, the variable ranges from 0 to 100, such that a coalition with 51% of its members allied will be 1.1 times more likely to break down than a coalition with 50% of its members allied. To get a better sense of the substantive meaning of the relationship, Figure 5.5 plots the hazard function for coalitions that are 25%, 50%, and 100% allied, across which the differences in the hazard are quite dramatic: coalitions with few allied members are substantially less likely to break down at a given time than those with even half of their members sharing alliances.

With Diversity measuring the degree of shared preferences in the coalition, Percent Allied captures all those effects of alliances not tied to similar preferences, which point to more formally committed coalitions being associated with more fragile post-victory cooperation. If alliances represent particularly costly forms of commitment (Morrow 1994, 2000), especially relative to informal pledges of cooperation, then Percent Allied may capture the underlying cooperative problems that necessitated the formation of the alliance in the first place. It is also possible, as well, that coalitions with a larger fraction of allies are comparatively more willing to take on partners – as the Western Allies were with respect to the Soviet Union – that will be unreliable after the war, because the allies expect to be able to work together in defense of the postwar

---

[25] However, it is also worth noting that drawing from a larger pool of states that might fall into war with one another increases the chances that some pair from the pool will do so. Absent additional controls for this baseline probability of war induced by larger group size, any specific claims about the size principle as a theoretical variable are problematic; as a control variable, Number captures both the size principle and the issue of group size, which is sufficient for my purposes here.

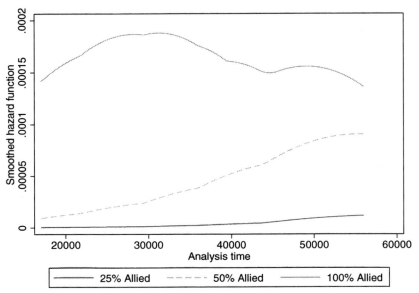

FIGURE 5.5. Hazard of coalitional breakdown by percent of states sharing ATOP Alliances, 1859–2007

status quo against unreliable partners. However, the average survival time for nonallied dyads is roughly 20,000 days by the COW measure and 19,000 for ATOP, while the average survival time for nonallied dyads is roughly 11,000 and 13,000 for COW and ATOP, respectively – suggesting that it is indeed allied states that turn against one another in competition over the spoils of victory.

What can we learn from this analysis? First, the fact that both preference diversity and the number of members achieve statistical significance shows that the effects of diversity are robust to variation in the size of the coalition, meaning that both numbers and preference diversity have independent effects. In other words, large coalitions are not vulnerable to breaking down in war because they are more likely to exhibit diverse preferences, nor are diverse coalitions vulnerable simply because they might also tend to be large. On the one hand, this is consistent with Riker's (1962) prediction that winning coalitions made overlarge by victory will tend to dissolve, and on the other, it indicates that diverse preferences do have an effect on their own that differs fundamentally from the size principle. Thus, where most coalitional work focuses on problems of collective action and burden-sharing, following Olson (1965), the challenges of distributive politics among members, embodied in the size principle and the problem of preference diversity, may point the way to new discoveries about the relationships between military cooperation, war, and peace.

5.4 APPLICATION: PAKISTAN, IRAN, AND THE SUPERPOWERS

The theory presented in this chapter also offers a plausible account for the differing international responses to the decade-long Soviet intervention in Afghanistan that began in 1979 and the American-led Gulf War of 1991, as well as the response to the Iraq War of 2003. As noted at the beginning of the chapter, a superpower led a military coalition against a weaker target in all three cases, yet while Pakistan intervened on behalf of the *mujaheddin* resistance during the Afghan War, Iran remained neutral during Operation Desert Storm in 1991. Further, while it did not intervene directly in the Iraq War of 2003 and its aftermath, Iran did provide support for militants that resisted the American occupation and the establishment of a new Iraqi government in the years after the invasion. Each conflict occurred in the same geographic neighborhood, involved a superpower as part of a militarily powerful coalition, and involved a number of similar actors. Yet why did Pakistan intervene in the early 1980s and Iran remain neutral in 1991, only to intervene against the United States-led coalition after 2003?

While all three coalitions were quite powerful militarily, a superpower victory in the Soviet Union's Afghan War would imply a durable aggregation of power that Pakistan would find difficult to accept:

Worried about the possibility of facing a coordinated attack from Afghan and Soviet troops on one front and the Indian military on the other, the first goal of Pakistan's military planners was the removal of Soviet forces from Afghanistan. *Even if an armed communist invasion of Pakistan was a remote possibility, intimidation was not.* There was also deep concern within the Pakistani government that Moscow would instigate, through material support, ethnic separatist movements in Baluchistan and the North-West Frontier Province. (Weinbaum 1991b, p. 72, emphasis added)

In 1979, before the repudiation of the Brezhnev Doctrine's promise of support for revolutionary movements around the globe, Soviet intervention in foreign countries was easily interpreted as a threat to others on its periphery, because the USSR's foreign policy had been for decades (at least rhetorically) aimed at tilting the global balance of ideology in favor of communism (see Braumoeller 2012, ch. 4). Further, victory in Afghanistan would have meant the preservation of its communist government, which would be dependent on the USSR for aid and guidance, as well as the long-term presence of Soviet troops, all part of a potentially very durable Soviet-Afghan coalition – one that would surely be more powerful for having secured a joint victory over the insurgency.

This possibility, combined with Russia's long-standing desire to acquire a warm-water port, set off alarm bells in the Pakistani government (Weinbaum 1991a, p. 496). Therefore, even as it devoted a substantial amount of its attention to its long-simmering and often violent rivalry with India to the east, Pakistan had strong incentives to prevent a communist victory, prompting it to spearhead opposition to the war by intervening on behalf of the Soviet-Afghan coalition's target: the *mujaheddin* (*ibid.*, pp. 497–499). Notably, and consistent with the theoretical model, Pakistan did not need to fear an explicit invasion – merely the continuation of Soviet-Afghan cooperation backed up by the coalition's aggregate military power, which could lead to substantial changes in the regional status quo even without war. As a result, Pakistan under General Zia ul-Haq went on to coordinate robust support for the Afghan resistance, providing cash and weapons (often from the United States and friendly Gulf countries), as well as basing and logistical support, dramatically increasing the Soviets' costs for maintaining the war effort (Amin 2010, Coll 2004, Kalinovsky 2011).

While Pakistan's support for the *mujaheddin* does satisfy my definition of military cooperation – it effectively provided basing and staging rights on its side of the Afghan border – its choice over calibrating its contribution is interesting in its own right. In the theoretical model, the third party can be deterred from balancing when the opportunity costs of doing so $(1 - \delta)$ are too high, because it would waste resources that could otherwise be used in its subsequent defense. However, third parties can generally *choose* their opportunity costs by limiting exposure to them, and Pakistan seems to have done just that: faced with two intervention options, one involving high opportunity costs (facing the Soviet and Afghan government forces on the battlefield) and one involving lower opportunity costs (supporting and sheltering cross-border insurgents), it opted for the latter. Fighting the Soviet military on the ground would have been expensive and dangerous, especially given Pakistan's own concerns about India, and less attractive than remaining on the sidelines. However, since more limited support could be given at lower opportunity costs, Pakistan still found a way to join the war and balance against a future threat.

The two American-led wars against Iraq can also be understood in light of the theoretical model. In contrast to the Soviet-Afghan coalition, the diversity of interests in the 1991 Gulf War coalition meant that, despite its power, it would likely be unable to remain intact to pose a collective threat to Iran. This led the latter to decide that the costs of opposition were unlikely to be justified. In the months leading up to the 1991 Gulf War, the United States built a diverse coalition that included nontrivial contributions from regional powers like Saudi Arabia, Egypt, Turkey, and Syria.[26] Israel, a close regional friend of the United States, was notable mainly for its absence, which is consistent with the logic of partner choice described earlier; a partner too likely to provoke balancing is simply unattractive, especially when its military contribution is not likely to be decisive. Thus, the participation of states with frequently dissimilar policy preferences, not to mention Saudi Arabia choosing to condition its participation on the limitation of coalition war aims (see Chapter 3), made it unlikely that the coalition would remain intact beyond accomplishing its goals in Iraq. Despite more than a decade of hostile relations and little reason to trust

---

[26] The United States also had the tacit support of the Soviet Union, and, as discussed further below, the explicit authorization of the UN Security Council.

the United States, Iran caused few problems for coalition forces during the war, maintaining neutrality even as a hostile superpower and its Sunni rivals in the region threatened to shift the regional balance further against Iran.[27]

By 2003, however, as the United States sought to replace Iraq's Ba'ath government, the military coalition was substantially less diverse. Saudi Arabia and Turkey each refused to join (see Chapter 3), and in addition to the United States, only the United Kingdom contributed to the invasion force, while Kuwait provided the staging ground. As a result, promises to disband the coalition after a successful invasion were likely incredible, because there were fewer partners to restrict subsequent American aims, as Saudi Arabia was able to do in 1991. In the shadow of American rhetoric in the months leading up to the war, typified in its most strident form by Iran's inclusion in what President George W. Bush called an "Axis of Evil" in his 2002 State of the Union address (Bush 2002), there existed substantial uncertainty over the extent of American aims in the Middle East (Remnick 2003). Faced with the prospect of future confrontations with a coalition rhetorically committed to transforming the politics of the region, the model suggests that Iran might have found opposition to coalitional efforts in the 2003 Iraq War attractive, and indeed media and government reports pointed to Iranian material support for the Iraqi insurgency throughout the conflict (Kilcullen 2006). Of course, Iran appears to have remained neutral during the invasion of Iraq, only increasing its involvement when it had the opportunity to support the insurgency. As discussed in Pakistan's case, this is consistent with the logic of the theoretical model, as overt opposition to the war might have risked too much – that is, δ was too small – but once Iran saw the opportunity to oppose the coalition more or less covertly, it took the opportunity to do so.

If the key distinction between the 1991 and 2003 coalitions is the diversity of their interests, then in the absence of documentary evidence about Iranian motivations, we might expect to observe differences in revealed preferences over foreign policy issues. Indeed, the variance of the 1990 UNGA ideal points for the members of the 1991 coalition is $\sigma^2 \approx 0.3678$, which we can compare this to the variance in ideal points of the members of the 2003 coalition, the United States, United Kingdom, and Kuwait, which is a substantially lower $\sigma^2 \approx 0.012$. I use data from 2000 for this calculation, as it is the most recent year for which data are available. Therefore, if the theory is correct, the 2003 coalition could make far less credible promises to disband than the 1991 coalition, which would have increased Iran's incentives to intervene.

While the rather large differences in preference diversity are consistent with the differences in Iranian responses to each war, there are two obvious alternative explanations to address. First, American aims in each war proved to be quite different, with the 1991 war focused on the removal of Iraqi forces

---

[27] In fact, after assenting to shelter a number of Iraq's most advanced fighter aircraft before the 1991 war, Iran later refused to return them (Atkinson 1993, p. 477).

from Kuwait and the degradation of Iraq's military (Bush and Scowcroft 1998, Gordon and Trainor 2006) and the second, of course, on toppling its government (Keegan 2005). It is plausible that the higher stakes in the 2003 war caused more concern in Iran than the rather limited aims of 1991 did. However, the American-Iranian rivalry was in the background of both wars and would have contributed to expectations over future crises in both cases (see Colaresi, Rasler, and Thompson 2007), regardless of coalitional aims in Iraq. Thus, differences in war aims need not pose a problem for the model's explanation of Iranian behavior; in fact, the lack of restraint imposed on those aims by pliable coalition members is consistent with both the theory in this chapter and the committed partner equilibria in the crisis bargaining model of Chapter 4.

Second, it is possible that the lack of broad international support that undermined American coalition-building in the 2003 Iraq War might have also encouraged third parties to feel safe in their opposition to the coalition's effort. The associated counterfactual would be that support from other great powers might have signaled that few other countries would join in opposition (Thompson 2006, Voeten 2005), thereby deterring Iran from doing so. However, while not fundamentally incompatible with both the theoretical model and the empirical models in Table 5.1, this alternative does not account for the less-than-overt nature of Iran's intervention which, while risky, was clearly calibrated to risk as little as possible in the event that its attempts to undermine the coalition's effort failed and the United States and its partners did, in fact, eventually turn to the east. As suggested earlier, Iran proved willing to oppose the coalition when it could do so at a level that would allow it to minimize its opportunity costs in the event of an ultimate coalitional victory, that is, when δ remained sufficiently high. Thus, while both the multilateral approval and coalitional durability arguments are consistent with Iranian action in 2003, the latter explains the limited nature of its intervention; nonetheless, additional documentary evidence, particularly from Iran, may be necessary to resolve the questions about this case satisfactorily.

5.5 SUMMARY AND DISCUSSION

Military coalitions, for all their coercive benefits, represent aggregations of power that may arouse fears of future exploitation in third-party states. Nonbelligerents are quite often concerned with the aims that coalitions cannot promise not to pursue in the event of victory in today's conflict, as highlighted by the fears expressed by Ottoman foreign minister Mehmed Talaat Pasha in justifying his country's alignment with the Central Powers in the First World War:

If we stayed neutral, whichever side won would surely punish Turkey for not having joined them, and would satisfy their territorial ambitions at our expense. (quoted in Horne and Austin 1930, p. 398)

Especially worried that Russia, Great Britain, and France would be unable to promise not to dismember the empire and strip it of both the Turkish Straits and its holdings in the Middle East, the Ottoman Empire balanced against the Allies in hopes of staving off their ultimate victory. In fact, such a division of the Empire's territories was central to Allied war plans (Philpott 2014, pp. 69, 70), but parts of the agreements were obviated by the Russian Revolution and the Soviet exit from the war. Fears of this sort notwithstanding, most coalitions in the latter half of the twentieth century were able to keep their targets isolated; however, nearly 20% of them provoked the formation of counter-coalitions or the intervention of other states on the enemy's side in crises. To explain this variation – and to clarify when the ostensible benefits of military cooperation might be undermined by the expansion of the conflict – this chapter presented a game-theoretic model centered on coalition formation, a third party's decision to intervene in an ongoing conflict, and a coalition member's decision to stay or leave after the war. The third party finds the coalition's members more threatening together than they would be separately, but it is uncertain over the partner's willingness to remain in the coalition beyond today's conflict, while the lead state would like, all else being equal, to build coalitions that are both durable and unlikely to provoke balancing.

In equilibrium, the third party uses revealed policy preferences inside the coalition to judge the likelihood that it will remain intact or disband. Intervening on the target's behalf lets the third party influence the coalition's chances of victory, its future bargaining strength, and, interestingly, its post-conflict durability. Weighing the coalition's military power and durability against the costs of intervention, an observer is less likely to intervene against a powerful coalition as its members' revealed preferences grow more diverse. However, when the coalition is weaker, the observer is more likely to intervene as its members' revealed preferences grow more diverse. Singletons, on the other hand are unconditionally more likely to provoke opposition the more likely they are to defeat their targets, and lead states take these factors into account when choosing partners, even to the point of turning down valuable contributions when they are especially likely to provoke balancing. An empirical model of coalitions and conflict expansion in interstate crises during the period 1946–2001 supports these expectations, offering a rationale for understanding how observers evaluate the threats posed by coalitions as opposed to individual states. Further, a separate analysis of the durability of war-winning coalitions is consistent with the theory's other prediction that diverse coalitions break down into intramural conflict sooner after military victory than more homogeneous coalitions do. Thus, coalitional politics – specifically, the choices made at the stage of coalition formation – can have lasting effects on patterns of war and peace even after the initial impetus for military cooperation has passed. The model can also provide a plausible account for the logic of decisions over balancing and remaining neutral in the Soviet-Afghan War, the Gulf War of 1991, and the Iraq War of 2003.

Although its empirical focus deals with the specific context of interstate crises, the theoretical model is also relevant for broader questions of how states identify and respond to potential threats. In particular, states apparently respond not to power alone but also to their beliefs about future intentions; to be sure, increasing power in the absence of coalitional diversity does raise the probability of balancing, but a diversity of preferences can substantially reduce it. Indeed, while Vasquez (2009) claims that "the attempt to use the device of an alliance to increase relative military power usually fails because of the formation of a counter-alliance" (p. 182), this does not appear to be the case for military coalitions in general; counter-coalitions do not appear always to be the most "appropriate" response to others' military cooperation. In fact, there appears to be no systematic relationship between the percent of allied pairs in coalitions and the provocation of balancing. While some powerful states are more threatening than others, so too are some coalitions more threatening than others; and neither singletons nor coalitions are unconditionally more likely to provoke balancing than the other. However, judging the level of threat posed by coalitions is different than judging that posed by individual states, since it involves calculations of the likelihood that they will disband after the current conflict or stay together and use their aggregated power in subsequent crises; future intent and future power are both implicated in the coalition's subsequent choice to disband. I examine one possible source of information about coalitional durability in the form of preference diversity as revealed through UNGA voting, though other indicators are plausible and may facilitate further refinement to the relationships explored here.

Finally, theories of balancing and alignment are also closely related to the broader question of war expansion. That is, why do most wars remain small while others see multiple states choosing sides and engaging in hostilities, even when the issues over which the war started appear to be of little direct relevance to the extra belligerents? Considerations of relative power (Altfeld and Bueno de Mesquita 1979), geography and alliances (Siverson and Starr 1990), and opportunistic predation (Gartner and Siverson 1996) have all been linked to the decision to enter someone else's war with the chief aim of gaining directly from the outcome of the current conflict. The theory and evidence presented here, however, suggest that the desire to balance against potential future threats can explain not only why some wars expand but also why those that already begin large – coalitional wars – can touch off additional rounds of expansion when fearful states face coalitions unlikely to disband after today's war. Every world war, after all, begins as something much smaller.

This chapter closes the discussion of the primary stages of military multilateralism: (1) coalition formation, (2) crisis bargaining, and (3) coalition expansion (and coalitional durability). In so doing, it has shown that the interaction of power and preferences can explain several patterns originally uncovered in Chapter 2 and, perhaps more importantly, that coalitions are not merely the sum of their parts, if we take "parts" to mean their aggregate military capabilities. Rather, the diversity of interests within them shapes the compensation required to ensure cooperation, the bargaining strategies used against targets, and, finally, the expansion of conflicts to include other states. However, the dialogue between theory and evidence need not end here. Each chapter raises additional questions about the broader process of military multilateralism, and these outstanding questions and possible extensions of the theory will figure prominently in the following chapter.

5.6 APPENDIX

5.6.1 Proofs

Proof of Proposition 5.1. The unique PBE exists when $\alpha < \alpha^{\dagger}$ and $\delta > \delta^{\dagger}$. Strategies are as follows. L: build iff $s < s^{\dagger}$. B: if L builds, balance iff $c_B < \hat{c}_B$; and if L does not build, balance iff $c_B \leq \tilde{c}_B$. P: if B balances, leave iff $c_P < \hat{c}_P$; and if B remains neutral, stay iff $c_P < \underline{c}_P$. Beliefs are as follows. If B balances after L builds, the other players believe $c_B \sim \text{U}(0, \tilde{c}_B)$ and $c_B \sim \text{U}[\tilde{c}_B, \overline{c}_B]$ if B remains neutral; if B balances after L acts unilaterally, the other players believe $c_B \sim \text{U}(0, \hat{c}_B)$ and $c_B \sim \text{U}[\hat{c}_B, \overline{c}_B]$ if B remains neutral. If P stays after B balances, the other players believe $c_P \sim \text{U}(0, \hat{c}_P)$ and $c_P \sim \text{U}[\hat{c}_P, \overline{c}_P]$ if P leaves; if P stays after B remains neutral, the other players believe $c_P \sim \text{U}(0, \underline{c}_P)$ and $c_P \sim \text{U}[\underline{c}_P, \overline{c}_P]$ if P leaves. Since each action occurs in equilibrium, there are no out-of-equilibrium beliefs. Further, since no player's payoffs depend on another's beliefs over its type, there are no incentives to against against type, ensuring that posterior beliefs are trivial; as such, consistency is assured, and I focus on establishing sequential rationality.

To verify that strategies are sequentially rational, begin with P's choice over staying in and leaving the coalition. For an arbitrary $I = \{0, 1\}$, P stays when

$$\beta b_{LP}(I) - c_P > \beta b_L(I),$$

or when $c_P < \beta(b_{LP}(I) - b_L(I))$. This establishes two cutpoints over P's type at which it is indifferent over staying and leaving,

$$c_P < \beta(b_{LP}(1) - b_L(1)) \equiv \underline{c}_P \quad \text{and} \quad c_P < \beta(b_{LP}(0) - b_L(0)) \equiv \hat{c}_P,$$

given intervention ($I = 1$) and neutrality ($I = 0$), respectively; otherwise, all higher types leave the coalition.

Next, consider B's decision over balancing and remaining neutral. If L does not build, then B balances when

$$\underline{p}_L(1 - b_L(1)) + \left(1 - \underline{p}_L\right) - c_B > \overline{p}_L(1 - b_L(0)) + (1 - \overline{p}_L),$$

or when $c_B < \overline{p}_L b_L(0) - \underline{p}_L b_L(1) \equiv \hat{c}_B$; otherwise, all higher types remain neutral. If L builds, then B's decision depends on P's likely response. As such B's expected utility for balancing is

$$\int_0^{\hat{c}_P} \left(\underline{p}_{LP} (1 - b_{LP}(1)) + \left(1 - \underline{p}_{LP}\right) \right) dU dc_P$$

$$+ \int_{\hat{c}_P}^{\bar{c}_P} \left(\underline{p}_{LP} (1 - b_L(1)) + \left(1 - \underline{p}_{LP}\right) \right) dU dc_P - c_B,$$

and its expected utility for remaining neutral is

$$\int_0^{\underline{c}_P} \left(\underline{p}_{LP} (1 - b_{LP}(0)) + \left(1 - \underline{p}_{LP}\right) \right) dU dc_P$$

$$+ \int_{\underline{c}_P}^{\bar{c}_P} \left(\underline{p}_{LP} (1 - b_L(0)) + \left(1 - \underline{p}_{LP}\right) \right) dU dc_P.$$

B balances when the former is strictly greater then the latter, or when

$$c_B < \overline{p}_{LP} \left(\beta \, (b_{LP}(0) - b_L(0))^2 + \bar{c}_P b_L(0) \right)$$

$$- \underline{p}_{LP} \left(\beta \, (b_{LP}(1) - b_L(1))^2 + \bar{c}_P b_L(1) \right) . \equiv \tilde{c}_B$$

Otherwise, all higher types remain neutral.

Finally, consider L's choice over building a coalition and acting unilaterally. Its expected utility for acting as a singleton is

$$\int_0^{\hat{c}_B} \left(\underline{p}_L (1 + b_L(0)) \right) dU c_B + \int_{\hat{c}_B}^{\bar{c}_B} \left(\overline{p}_L (1 + b_L(0)) \right) dU dc_B,$$

while its expected utility for accepting P as a coalition partner is

$$-s + \int_0^{\tilde{c}_B} \left(\underline{p}_{LP} \left(1 + \int_0^{\underline{c}_P} b_{LP}(1) dU dc_P + \int_{\underline{c}_P}^{\bar{c}_P} b_L(1) dU dc_P \right) \right) dU dc_B$$

$$+ \int_{\underline{p}_{LP}}^{\bar{c}_P} \left(\overline{p}_{LP} \left(1 + \int_0^{\hat{c}_P} b_{LP}(0) dU dc_P + \int_{\hat{c}_P}^{\bar{c}_P} b_L(0) dU dc_P \right) \right) dU dc_B.$$

L builds a coalition when the latter is strictly greater than the former, or when

$$s < (b_L(0) + 1) \overline{p}_L (\tilde{c}_B - \bar{c}_B) + \overline{p}_{LP} (\bar{c}_B - \hat{c}_B)(b_L(0) \bar{c}_P + \hat{c}_P(b_{LP}(0) - b_L(0)) + 1)$$

$$- (b_L(1) + 1) \tilde{c}_B \underline{p}_L + \hat{c}_B \underline{p}_{LP} (b_L(1) \bar{c}_P + \underline{c}_P(b_{LP}(1) - b_L(1)) + 1) \equiv s^\dagger,$$

and acts unilaterally otherwise. Therefore, the proposed strategies constitute the unique Perfect Bayesian Equilibrium. □

Proof of Proposition 5.2. To verify the claim, recall that the probability that B is of a type that balances is $\Pr(c_B < \tilde{c}_B) = \tilde{c}/\overline{c}_B$, and

$$\frac{\partial \Pr(c_B < \tilde{c}_B)}{\partial \overline{p}_L} = \frac{b_L(0)}{\overline{c}_B} > 0,$$

which is strictly positive. □

Proof of Proposition 5.3. To verify the claim, let $c_P < \beta (b_{LP}(I) - b_L(I)) \equiv c_P^{\dagger}$ denote the player-types of P that stay for any $I = 0, 1$. Since $\Pr(c_P < c_P^{\dagger}) = c_P^{\dagger}/\overline{c}_P$, we have

$$\frac{\partial (c_P^{\dagger}/\overline{c}_P)}{\partial \beta} = \frac{b_{LP}(I) - b_L(I)}{\overline{c}_P} > 0,$$

which is strictly positive for any $I = 0, 1$. □

Proof of Proposition 5.4. To verify that $\Pr(c_P \leq \underline{c}_P) < \Pr(c_P \leq \hat{c}_P)$, we need to establish first that $\hat{c}_P/\overline{c}_P > \underline{c}_P/\overline{c}_P$, which is true when

$$b_{LP}(0) - b_L(0) > b_{LP}(1) - b_L(0).$$

Then, replacing the reduced form parameters with their components as defined in Equation (5.1) and resolving the inequality in terms of α and δ shows that the inequality can be satisfied under several sets of conditions. First, when $m_L \geq m_B$, the inequality is satisfied when

$$\alpha < \frac{m_B^2}{m_L(m_L + m_P)} \quad \text{and} \quad \delta > \frac{\alpha m_L(m_L + m_P)}{m_B^2}.$$

Second, when $m_L < m_B$ and $m_P < (m_B^2 - m_L^2)/m_L$, the inequality is satisfied when $\alpha < \delta$ and

$$\delta \geq \frac{m_L(m_L + m_P)}{m_B^2}.$$

Finally, when $m_L < m_B$ and $m_P \geq (m_B^2 - m_L^2)/m_L$, the inequality is satisfied for any δ when

$$\alpha < \frac{\delta m_B^2}{m_L(m_L + m_P)}.$$

When any of these sets of conditions are met, $\Pr(c_P \leq \underline{c}_P) < \Pr(c_P \leq \hat{c}_P)$. □

Proof of Proposition 5.5. To verify this claim, note that the first partial derivative of $\hat{c}_B$ with respect to β is

$$\frac{\partial \hat{c}_B}{\partial \beta} = \overline{p}_{LP}(b_{LP}(0) - b_L(0))^2 - \underline{p}_{LP}(b_{LP}(1) - b_L(1))^2,$$

which is positive when

$$\overline{p}_{LP} > \underline{p}_{LP} \frac{(b_{LP}(1) - b_L(1))^2}{(b_{LP}(0) - b_L(0))^2} \equiv \overline{p}_{LP}^{\dagger},$$

and weakly negative when $\overline{p}_{LP} \leq \overline{p}_{LP}^{\dagger}$. $\qquad\qquad\square$

Proof of Proposition 5.6. To verify the claim, we consider $s^{\dagger}$ in reduced form, using the expected utilites in the proof of Proposition 5.1. Since the probability of balancing is $\hat{c}_B/\overline{c}_B$, we take the first partial derivative with respect to $\hat{c}_B$,

$$\frac{\partial s_{LP}}{\partial \hat{c}_B} = \underline{p}_L \left(1 + \underline{c}_P[b_{LP}(1) - b_L(1)] + \overline{c}_P b_L(1)\right)$$

$$-\overline{p}_{LP}\left(1 + \hat{c}_P[b_{LP}(0) - b_L(0)] + \overline{c}_P b_L(0)\right) < 0,$$

which is sure to be negative, since $\overline{p}_{LP} > \underline{p}_L$; $\underline{c}_P < \hat{c}_P$ and $b_{LP}(0) - b_L(0) > b_{LP}(1) - b_L(0)$, by Proposition 5.4; and $b_L(0) > b_L(1)$. $\qquad\square$

5.6.2 Victorious Coalitions and their Members

Table 5.3 lists the coalitions and their members that COW identifies as winning the relevant conflict (Sarkees and Wayman 2010). I generated the list by eliminating all bilateral wars from the sample, as well as those in which singletons defeat coalitions. The remaining conflicts are those in which coalitions won the war, as opposed to losing or seeing the war end in stalemate.[28] In some cases, wars transition from interstate to another type of war, often intrastate war. In these cases, if the interstate phase had a clear winner – like the Iraq War, in which the coalition successfully toppled the target government before the transition to civil war – I coded the war's end at the transition point and the outcome accordingly. This leads to the inclusion of the War Over Angola (COW #186), Phase Two of the Second Ogaden War (COW #187), the 2001 Invasion of Afghanistan (COW #225), and the 2003 Iraq War (COW #227).

---

[28] See Starr (1972) for an analysis of the distribution of payoffs between both winning and losing coalitions.

TABLE 5.3. *Victorious Coalitions and Member States*

COW #	War	End Year	Coalition	Balance
16	Roman Republic	1849	France Austria Two Sicilies	
22	Crimean	1856	United Kingdom France Italy Turkey	
28	Italian Unification	1859	France Sardinia-Piedmont	
46	Second Schleswig- Holstein	1864	Germany Austria	
49	Lopez War	1870	Brazil Argentina	
52	Naval War	1866	Peru Chile	
55	Seven Weeks' War	1866	Prussia Meck/Schwerin Italy	Yes
58	Franco-Prussian	1871	Bavaria Prussia Baden Wurttemburg	
82	Boxer Rebellion	1900	United States United Kingdom France Russia Japan	
100	First Balkan	1912	Serbia Greece Bulgaria	
103	Second Balkan	1913	Serbia Greece Romania Turkey	

(*continued*)

TABLE 5.3 *(continued)*

COW #	War	End Year	Coalition	Balance
106	World War I	1918	United Sates United Kingdom Belgium France Portugal Italy Serbia Greece Romania Russia Japan	Yes
107	Estonian Liberation	1920	Estonia Finland	
108	Latvian Liberation	1920	Estonia Latvia	Yes
112	Hungarian Adversaries	1919	Czechoslovakia Romania	
136	Nomonhan War	1939	Soviet Union Mongolia	
139	World War II	1945	United States Canada Brazil United Kingdom Netherlands Belgium France Poland Italy Yugoslavia Greece Bulgaria Romania Soviet Union Norway Ethiopia South Africa China Mongolia Australia New Zealand	Yes
155	Sinai War	1956	United Kingdom France Israel	

COW #	War	End Year	Coalition	Balance
158	Ifni War	1958	France Spain	
186	Angola	1976	Cuba Angola	Yes
187	Second Ogaden, Phase II	1978	Cuba Ethiopia	
211	Gulf War	1991	United States Canada United Kingdom France Italy Morocco Egypt Syria Saudi Arabia Kuwait Qatar United Arab Emirates Oman	
221	Kosovo	1999	United States United Kingdom Netherlands France Germany Italy Turkey	
225	Invasion of Afghanistan	2001	United States Canada United Kingdom France Australia	
227	Invasion of Iraq	2003	United States United Kingdom Australia	

6

Conclusion

> Probably the essential fact about the role of leader in a world-dominating coalition is that the actor who once occupies this role cannot resign from it without unpleasant consequences.
>
> William H. Riker
> *The Theory of Political Coalitions*, p. 219

This book opened with a question about the role of military cooperation in explaining patterns of war and peace in the international system. Despite a long tradition of research into the formation and behavior of formal military alliances, international relations scholarship has been curiously silent on what is actually the dominant mode of observed military cooperation: the coalition. Treating coalitions as part of a general process of military multilateralism, whereby states seek out and secure military contributions during international crises, the approach developed in the preceding chapters shows that two distributions – (a) military power across disputing sides and (b) foreign policy preferences within coalitions – interact to explain a range of outcomes of enduring interest to students of international relations. Many of these questions, I have argued, are at best only partially understood without reference to intra-coalitional politics, or the "diplomacy of co-belligerency" (Fowler 1969, p. 4). Within this framework, I develop a simple transactional theory of military cooperation, where costly contributions to a coalition's coercive efforts require some reciprocal amount of costly compensation. This allows the core theoretical and empirical chapters to explore the linkages between the choice of coalition partner, the escalation of crises to war, and the expansion of conflicts beyond their original participants.

After Chapter 2 established military coalitions as phenomena distinct from both formal alliances and instances of diplomatic multilateralism, the subsequent chapters went on to develop and empirically assess the following sets of expectations:

- **Coalition formation (Chapter 3):** States prefer coalition partners with preferences similar to their own in order to minimize compensation payments, but they become less selective the more militarily valuable a given partner will be.
- **Crisis escalation (Chapter 4):** Coalition partners that fear an overly costly war can raise the probability of the very war they fear against weak targets, but they lower the probability of war against stronger targets.
- **Conflict expansion (Chapter 5):** An increasing diversity of member preferences discourages observers from balancing against strong coalitions, but diversity actually encourages balancing against weaker coalitions.

Leveraging the insights generated about coalition formation, which explain the occurrence of diverse coalitions through an interaction of military contributions and required compensation, the chapters on escalation and expansion show that intra-coalitional politics plays a substantial, and heretofore unappreciated, role in both processes. For example, the conventional wisdom that coalition partners pose problems for sending credible signals of resolve appears incorrect; at times, a partner's desire to restrain the coalition leader's bargaining position can leave information problems unsolved and generate a risk of war – a war that would not happen if cooperation were not possible – but at others it discourages risky bluffing and facilitates peace. Likewise, a diversity of preferences within coalitions discourages balancing against powerful coalitions that would otherwise have difficulty promising restraint, while the same diversity can actually encourage balancing and conflict expansion against weaker coalitions. Further, each chapter shows that preference diversity can account for differences in the rates of war and conflict expansion across coalitions and equally powerful singletons, pointing the way to a richer set of comparisons than those conducted in Chapter 2.

While the models above were developed with an eye to explaining specific empirical patterns uncovered in Chapter 2, they also identify causal processes and mechanisms that can be used to shed light on a number of different questions, including the durability of cooperation between coalition members after victory in war – which was analyzed in detail in Chapter 5. In addition to showing that diverse coalitions break down into violence over the peace just secured more quickly than homogeneous coalitions do, the empirical analysis also showed that those coalitions made up of a high proportion of allied states are also uniquely likely to break down in intramural violence. This result is especially interesting in that it points to some potentially surprising differences between allied and nonallied coalitions. The survival and breakdown of victorious coalitions is central to understanding the survival of world orders built after systemic wars, as well as to the basic politics of "winning the peace" after smaller-scale coalitional conflicts, and Chapter 5 offers several new avenues of inquiry into this problem.

In the rest of this chapter, I explore two other possible extensions of the approach developed here. The first follows directly from the model of coalition formation analyzed in Chapter 3: how the nature of the stakes of an international crisis shapes both the possibility of cooperation and coalitional crisis bargaining. Second, I leverage the theories and empirics of Chapters 4 and 5 to consider future research on the differences between allied and nonallied coalitions, particularly how formal commitments might alter the intra-coalitional bargaining processes highlighted throughout the book. Then, I summarize the implications generated by the dialogue to this point for the United States, the most prominent contemporary leader of military coalitions, as it enters a period marked by both substantial military superiority and uncertainty over the timing and nature of potential future power transitions. Finally, I conclude with a discussion of how future work on the topic military coalitions might use the theory's core insights to continue the exchange between puzzles, theory, and evidence – the modeling dialogue (see Myerson 1992, Powell 1999) – begun in this volume.

6.1 PUBLIC AND PRIVATE STAKES IN INTERNATIONAL COOPERATION

The value of a theory or theoretical approach consists of not only how well it answers the questions that inspired it but also what else we can learn from it – what other puzzles it resolves or what other predictions it makes about related phenomena. Indeed, while the empirical models in Chapters 3–5 above are focused on the specific outcomes of coalition formation, crisis escalation, conflict expansion, and coalitional durability, the theoretical models made several other predictions about related processes, and in this section I discuss one of them in particular: how the public or private nature of the stakes of crises shapes incentives for military cooperation in the first place.

Chapter 3 indicates the presence of a link between (a) the relative size of the public and private components of the stakes and (b) a lead state's ability to attract support when it desires a coalition partner. Where most scholarship associates an increasing public goods component with decreased levels of cooperation (cf. Olson 1965, Olson and Zeckhauser 1966), this relationship – at least in terms of military cooperation – depends critically on a potential partner's military contribution. Specifically, an increasing public component makes coalitions more likely with weaker partners, who provide a smaller military boost, and less likely with stronger ones, who provide more substantial military benefits. As a result, states might often bring along smaller powers for public stakes and larger powers for conflicts with a larger relative share of spoils to distribute, be they conquered lands, control over defeated governments, or indemnities.

The second pattern is initially counterintuitive. Why take on more powerful partners, who can leverage the threat not to participate in order to capture a larger share of the spoils in return for cooperation, precisely when those spoils account for relatively more of the outcome? The answer lies in a coalition builder's ability to shape the terms of a partner's cooperation. While, all else being equal, a powerful partner can demand larger side payments in return for cooperation, its threat to stay on the sidelines is also undermined by smaller public goods components, making it relatively more eager to join the coalition for smaller levels of compensation. On the other hand, if the partner is powerful and there is a large public component, securing its cooperation will be much more difficult. If military cooperation is fundamentally transactional, then thinking in terms of potential partners' outside options should be key to understanding the price at which their cooperation can be purchased (see also Fang and Ramsay 2010, Voeten 2001). In the case of coalitional action, the outside option depends not only on military power but also on the share of the outcome that states would get whether or not they participate – a fact overlooked in virtually all extant explorations of treaties of alliance, diplomatic multilateralism, or wartime coalitions.

How might we apply this insight? Straightforwardly, scholars can think in new ways about how to conceive of the variety of issues over which states contend. Traditional distinctions between territorial and nonterritorial issues (Holsti 1991, Senese and Vasquez 2008, Vasquez 2009) or security and economic issues (Keohane 1984, Keohane and Nye 1977), for example, may be inappropriate for identifying the relative sizes of public and private shares of a given international outcome. Starr (1972) explores the distribution of private payoffs after war, but public payoffs are of course more difficult to measure. However, as discussed in Chapter 3, interventions into ongoing conflicts designed to eliminate some public "bad," such as genocide, ethnic cleansing, refugee problems, or contagious civil conflict might well be identified with a relatively larger share of public goods: many states have an interest in seeing a resolution to the crisis whether or not they participate in solving it.[1] On the other hand, crises involving the acquisition or defense of territory, the exploitation of resources, the extraction of indemnities, or the political institutions of target states generate fewer public goods relative to the private (or club) gains that can be divided among members after victory.

The content of such issues might productively be measured on a crisis-by-crisis basis, much as other scholars have measured the issue content of militarized disputes or crises (e.g., Ghosn, Palmer, and Bremer 2004, Hensel et al. 2008, Holsti 1991, Wilkenfeld and Brecher 2010), and then related to

---

[1] A "public bad" is non-rival and non-excludable, yet it harms welfare, as opposed to a public good, which shares those characteristics but enhances welfare. Therefore, solving a public bad generates a public good, which activates the traditional notion of collective action problems in the search for a solution (cf. Olson 1965, Sandler 1992).

predictions over both the formation of coalitions and the choice of partner. Thus, where prior work linking the nature of issues to patterns of war and peace focuses on matters of salience, be it social or political (Holsti 1991, Vasquez 2009), or the connection of issues to the technology of bargaining and war (Carter 2010), the present theory suggests that another dimension – the public/private nature of the stakes – can affect the chances of war by first affecting the prospects for cooperation. More generally, the model highlights some limitations of the conventional approach to understanding international cooperation generally, which focuses on incentives to defect and threats of punishment where the terms of cooperation are exogenously given in some variant of the Prisoner's Dilemma (Carrubba 2005, Downs, Rocke, and Barsoom 1996, Jervis 1978, Keohane 1984, Snyder 1997); when states can offer side payments to alter the terms of cooperation, some of the conventional wisdom, particularly that cooperation over public goods is difficult to achieve absent threats of punishment, may need revision. Many analyses of international institutions and their role in facilitating cooperation, for example, abstract away from the possibility of side payments and compensation, but the present approach shows that these processes can have a substantial impact on whether and how states attempting to cooperate achieve their goals. Such insights, however, await both the generalization of the basic cooperation model of Chapter 3 to different substantive settings and renewed effort at opening up the content of international issues.

6.2 ALLIED AND NONALLIED COALITIONS

Chapters 4 and 5 draw explicit comparisons between singletons and coalitions, although the latter are chiefly distinguished according to the diversity of their preferences. However, coalitions also vary in the extent to which they are tied together by formal commitments, allowing for a distinction between allied and nonallied coalitions. The former threaten war together as the result of honoring prior commitments to do so, and while the latter may involve allied states, their cooperation is not mandated by treaty; like members that do not share alliance ties, they simply choose to cooperate militarily. Two questions follow immediately. First, how do allies making threats and fighting together differ from states doing so on the basis of informal or ad hoc commitments? Second, are allied coalitions more or less effective than nonallied coalitions at sustaining cooperation after winning coalitional wars? In this section, I argue that we can think of answering these questions by focusing on two factors: (a) the relative costs of refusing military cooperation inside and outside alliances and (b) the underlying cooperative problems that necessitate alliance commitments in the first place.

Alliances are contracts that stipulate wartime behaviors, from the contribution of military effort to the maintenance of neutrality (Gibler and Sarkees 2004, Leeds et al. 2002, Morrow 2000), and as such they are attempts to shape the construction and membership of coalitions for an uncertain future. One of the primary means through which treaties of alliance affect state decisions over joining coalitions is by raising the costs of refusing military cooperation (Gibler 2008, Morrow 1994) – that is, the costs of skittishness (cf. Wolford 2014b, p. 154). If alliances are more costly to violate than ad hoc or informal commitments, then they should encourage some partners to be committed where they would otherwise be skittish, rendering those partners more willing to accept large mobilizations that would otherwise drive them to withdraw from the coalition.[2] Since committed partners cannot limit a lead state's level of mobilization, they may raise the chances of war against strong targets, but they might mitigate problems of lead states choosing smaller mobilizations in order to keep the coalition together, lowering the chances of war against weak targets; with more firmly committed partners, resolute lead states should be less hesitant to reveal themselves out of a fear of forfeiting coalitional support.

While the empirical model in Chapter 4 finds no relationship between the percent of allied members in a coalition and the escalation of crises to war, some of the potential effects of alliance ties are more subtle. There should be no direct relationship between the presence of alliance ties in the coalition and the probability of war; whether a skittish partner's ability to restrain a lead state lowers the chances of war, relative to a situation in which refusing to cooperate is less credible, depends again on the strength of the target. Nonetheless, alliances vary widely in their terms, and it is possible that other provisions can more successfully induce restraint; in fact, Benson (2012) shows that many alliances are built around the very possibility that some members may wish to opt out when their partners are too aggressive. On the other hand, those states that build ambiguity into their alliances do so when their preferences diverge (*ibid.*, p. 138), undermining any possible reduction in skittishness in some cases. Therefore, while the general relationship between preference diversity, the distribution of power, and the onset of war is likely to emerge in both allied and nonallied coalitions, the conditions under which the terms of specific alliance commitments can exacerbate or mitigate them, relative to an absence of commitments between coalition members, await further exploration.

Next, the empirical models in Chapter 5 show that increasing shares of allied members are unrelated to coalitions' rates of provoking balancing, even as war-winning coalitions with a higher percentage of allies among their members tend to break down into intramural war sooner than those with fewer allies. In peacetime, alliances can bolster general deterrence (Leeds 2003b, Morrow 1994) and increase trade (Gowa and Mansfield 1993, 2004, Mansfield and Bronson 1997), as well as provide some degree of control over states that would otherwise pursue undesirable policies (Morrow 1991, Pressman 2008, Weitsman 2004). However, these benefits may be associated with the downstream cost of an increased risk of future conflict once allies eliminate

---

[2] This discussion borrows liberally from Wolford (2014b).

or subdue a common threat on the battlefield and the original motives for signing treaties of alliance have fallen away. To the extent that alliances are costly to form (Morrow 1994), states might only establish them when doing so will resolve underlying cooperative problems – problems that are particularly likely to reemerge once former partners must agree on shares of the spoils just captured in victory. Thus, where alliances might be expected to help solve the problems faced by successful winning coalitions by raising the costs of noncooperation, they are associated here with a more rapid disintegration of winning coalitions. Why this might be the case, as well as what types of alliances are more dangerous after victory than others – to say nothing of the durability of war-winning coalitions itself – is a promising avenue for further research as well.

Finally, this discussion has thus far relied on ideal types: allied and nonallied coalitions. In practice, however, we are likely to see a variety of configurations of allied and nonallied states joining coalitions together, and the proper way to measure these differences will be a challenge. In Chapters 4 and 5, I use a rough measure of alliance ties in the form of the percent of partner-pairs that share some form of alliance commitment, but this remains imperfect; many allies, after all, fight together in conflicts not covered in the terms of their treaty commitments. How should we view such partners as opposed to those fighting to honor commitments or those bound by no treaty at all? Finally, when do states choose allies as coalition partners, and when do they choose nonallied partners? Chapter 3 shows that allies are more likely to join lead states' coalitions than nonallies, but they rarely do so to the exclusion of nonallied states. Each of these questions, though well beyond the scope of this book, should shed additional light on the relations between allied and nonallied coalitions, both of which great powers like the United States have made use of in the past and likely will use again in the future.

6.3 THE UNITED STATES AND MILITARY MULTILATERALISM

The preceding chapters can also shed light on the behavior of and potential problems facing particular states as they navigate international crises. The United States, for example, emerged from Allied victory in the Second World War as the globe's dominant military and economic power, and it has maintained that position to varying degrees through the Cold War, the collapse of the Soviet Union, and the current "unipolar moment" (Krauthammer 1990), when it became the world's only superpower. It has survived in this position, to a great degree, by securing the assistance of military coalitions that, a century ago, were virtually unknown to its leaders.

Coalition diplomacy, novel to and grudgingly practiced by the United States [during the First World War] in 1917–18, within 25 years became an identifying and permanent characteristic of American foreign policy. (Fowler 1969, p. 7)

Further, as noted presciently by Riker (1962) in this chapter's epigraph, leading a global coalition in defense of a hard-won status quo is no small task, and "the actor who once occupies this role cannot resign from it without unpleasant consequences" (p. 219). The United States seems likely to retain its hegemonic status for some time (Chan 2005, 2008), which means taking the lead in building coalitions to preserve the status quo, such as the defense of South Korea since 1950, or to reverse attempted changes to it, such as the restoration of Kuwaiti sovereignty in 1991. If the United States is likely to continue in its role as chief status quo coalition-builder in the near future, then we might ask what the preceding analysis can tell us about the future of American military multilateralism. In this section, I offer some preliminary answers to this question, leveraging the insights presented earlier and, where the analysis raises additional questions, suggesting lines for future inquiry into the politics of great power military coalitions.

Before applying the book's theory of military multilateralism to the United States, it is necessary to see where the United States "fits" – that is, where it is likely to score in terms of the key variables – in the models from the previous chapters. First and foremost, the United States will nearly always find itself scoring high on measures of military capabilities, whether measured by forces-in-being and power projection ability, latent capabilities such as industrial capacity and access to credit, or some composite of the two (see, e.g., Schultz and Weingast 2003, Singer, Bremer, and Stuckey 1972). That means, on balance, that the United States will often be tempted to act unilaterally because it will find whatever assistance other countries can offer rarely worth the costs of securing their cooperation, just as it did when it made a minimal offer of compensation to Turkey ahead of the 2003 Iraq War, confident in its chances of victory if cooperation were not forthcoming and, as such, willing to risk the failure of its proposal (see Chapter 3). Indeed, much of the observed American propensity to build coalitions is due to its uniquely high level of military capabilities; powerful lead states appear uniquely likely to build coalitions only because of their unusually high rates of participation in crises. On the other hand, as a potential partner, the United States is also uniquely attractive militarily, and the same chapter also suggests that it is quite likely to join coalitions at the request of weaker lead states, which it does on several occasions in the coalitions data. In fact, re-estimating the coalition formation models in Table 3.2 with a dummy variable indicating whether the United States is a potential partner shows a statistically discernible increase in the probability of coalition formation.

Next, the United States is not only uniquely powerful but also significantly more active around the world than in its own neighborhood than other states, most of whom tend to confront only their neighbors.[3] Thus, while it can be

---

[3] For a decomposition analysis that separates the effects of aggregate power from the "major power" designation, see Chiba, Machain, and Reed (2014).

choosy over partners that offer some types of contributions – say, troops, air-craft, or ships – it often finds itself negotiating over basing or staging rights, as it did for the 1991 and 2003 wars against Iraq, the 2001 invasion of Afghanistan, and even the air campaign over Kosovo in 1999. In each case, ostensibly small countries of little value in terms of military capabilities proved critical for placing American forces close to the crisis or conflict zone, reducing the costs of each operation and ensuring that they would stand an acceptable chance of success. Counterfactually, an air campaign over Kosovo waged solely by carrier-borne aircraft and strategic bombers launched from the United States would have come at tremendous cost *and* limited coercive power. In other words, projecting power far from home is difficult, and when powerful coun-tries operate beyond their immediate geographical neighborhood – which the United States is uniquely likely to do – they find themselves requiring the assis-tance of coalition partners despite their own significant absolute capabilities. Thus, those countries best able to project power beyond their borders will find themselves far more often dependent on partners to exercise that power than are states with lesser capabilities.

In a clear case of a powerful country eschewing coalitional support, the United States acted alone (militarily) in the Haitian crisis of 1994, which it could do easily given the geographical proximity of the target country. However, in the case of Saudi Arabia during the Gulf War in 1991, a nominally small but militarily critical country was able to hold significant sway over American war aims, and in the future it is likely to be this kind of partner – small, close to a target country, and offering a particular contribution that few others can – that exercises the most leverage over the United States in intra-coalitional politics. On the other hand, powerful potential partners like France or the United Kingdom might, despite their own considerable military capabilities, prove less influential over coalitional politics because their contributions are redundant or, more accurately, less important than those contributions that guarantee easy access to and logistical support in a conflict zone. All of this suggests, perhaps counterintuitively, that *what* partners contribute, especially when they are the only potential provider, can be just as important as *how much* they contribute to the coalition. The models presented throughout the book assume that the partner's contribution is specific – that is, on offer from no other states – but extensions to the case of multiple potential partners, all of which offer similar military contributions, is an interesting avenue for further inquiry. In particular, the former should exercise far greater influence over intra-coalitional politics than the latter, as threats to revoke cooperation are more credible when the contribution cannot be purchased elsewhere.

When it comes to crisis bargaining, Chapter 4 shows that a partner's abil-ity to shape a coalition's bargaining position, particularly when the partner can credibly refuse to participate in an overly costly conflict, can affect the probability of war. In particular, skittish partners can discourage risky bluff-ing against powerful countries, which may be comforting in the context of high-stakes disputes with other great powers. However, the military value of skittish partners can also discourage large mobilizations and credible signaling against weak targets, making war more likely than it would be either against strong targets *or* in the absence of coalition partners in the first place. Given the recent infrequency of great power conflict, it might be reasonable, barring the rise of China or the spread of conflict into Russia's former imperial Near Abroad, to expect that most crises in which the United States becomes embroiled will involve distant, relatively weak targets for which basing and staging contributions will be critical. Like Saudi Arabia in 1991 and Germany/Italy in 1999, such partners near the conflict zone are also likely to be skittish, seeking to limit the costs of the war and, tragically, undermining credible signaling that would otherwise lower the chances of war. Additionally, the United States' most likely problem will involve signaling resolve – its valuation of the stakes or costs of fighting – because significant military disparities mean that martial effectiveness, about which it is difficult to send credible signals (Arena n.d.), will matter less than it would at more even distributions of power. This pattern would change, and change rapidly, were great power conflicts to make a comeback in international relations.

The challenge of securing the cooperation of skittish partners inverts the traditional problem of entrapment in the theory of alliances, where strong states worry about inducing risky behavior on the part of the small states they are pledged to defend (cf. Snyder 1984, 1997). While powerful countries do tend to design alliance contracts to reduce problems of moral hazard, creating ambiguity in order to make sure that their protégés do not shift too much risk onto the shoulders of their defenders (Benson 2012), the primary problem in building military coalitions is not one of the pre-crisis distribution of risk but the distribution of the costs of a looming war, which fall unevenly across coalition partners. How the United States deals with such partners, and whether compensation comes in the form of moderated mobilizations or war aims as opposed to direct side payments – which are unlikely to affect signaling in the same way – will do much to explain why, in some cases, United States–led coalitions see their crises resolved peacefully and why, in others, they fall into war.

As it weighs options in choosing partners, the United States faces a unique problem of reassurance: as the most militarily capable state in the international system, the states with which it deals – and those that watch these dealings from the sidelines – are often reasonably concerned about the possibility of expansionism and exploitation (Ikenberry 2001, Powell 1999, Voeten 2005, Waltz 1979). As Chapter 5 shows, this problem takes on extra dimensions when states act with coalition partners, particularly when observers not initially involved in an ongoing conflict fear that victorious coalitions will continue cooperating after the war, taking advantage of their newly powerful position to turn against fresh targets. However, since the United States will be involved in relatively powerful coalitions, thanks to its own military capabilities, it can reduce the chances of third-party balancing by choosing coalition partners with whom cooperation will be credibly short-term, while avoiding those with threateningly similar interests; an increasing diversity of preferences is associated with reduced chances of conflict expansion for strong coalitions. As noted earlier, the menu of choice for coalition partners is often quite limited, and the United States may nonetheless choose partners that are likely to provoke fears in observer states. Steps such as neutralizing Taiwan and Israel in 1950 and 1991, respectively, may not always be feasible. Awareness of the challenge of reassurance under these conditions, however, can point policy makers toward the need to take additional steps, such as UNSC approval (Chapman 2011, Voeten 2005), that might alleviate the fears of potential balancers or, at a minimum, encourage them to remain neutral. Nonetheless, Chapter 5 also shows that the larger coalitions made possible by diplomatic multilateralism can also provoke balancing themselves, especially as coalitions increase from two to three or four members; only the largest among them tend to see benefits in reducing the rate of balancing.

Many of the implications in this section depend on the assumption of continuing American military and economic dominance, as well as a general absence of conflict among the great powers. How might these expectations change if either, or more likely both, situations were to change? First, and foremost, if the reach of American interests remains the same while military and economic resources grow scarce in a process of absolute decline, we might expect a greater reliance on coalition partners.[4] While the United States is already prolific in its use of military coalitions (Kreps 2011, Tago 2007), it has shown a certain amount of selectivity – for example, foregoing numerous contributions to the Korean War, as well as Turkish cooperation against Iraq in 2003 – that it might be less able or less inclined to show the more the pursuit of its goals requires the support of coalition partners.[5] As shown in Chapter 4, a combination of decreased selectivity and militarily more valuable contributions might mean that the United States will more often find itself with necessary but skittish partners that moderate its threats. This, of course, may come with the cost of an increased chance of war against the relatively weak targets against which superpowers find themselves set against almost by definition. However, Chapter 5 suggests that an increased diversity of interests may also presage a reduced chance of provoking balancing coalitions, a gain to be weighed against the costs of reductions in available military resources.

Next, what if the modal crisis opponent of the United States changes from minor or middling powers to great powers, perhaps due to the rise of China and increasing tensions in the Asia-Pacific region or increased Russian assertiveness

---

[4] For an argument that demographic change may drive all the great powers into absolute decline, implying a reduction of foreign intervention in general, see Haas (2007).

[5] See Braumoeller (2012) for a discussion of how declining relative power or resources leads to increases in security-related activity by the great powers.

in Eastern Europe? The logic of coalition formation in Chapter 3 suggests that projecting power far from home and facing an opponent on more equal military footing should render the United States generally less selective over its choice of partner. This is not without precedent; the United States cooperated with the Soviet Union extensively in the Second World War. At the same time, however, potential partners might see their foreign policy preferences moving closer to the United States if they feel threatened by a rising China or resurgent Russia, which also makes coalition formation easier by increasing the range of compensation deals that potential partners might find acceptable. As a result, coalition-building might become easier as superpower politics grows more contentious, both because the United States will have a greater need of partners and because those partners will be increasingly willing to cooperate for ever smaller concessions. Though it lies outside the scope of the theory, a loosening of constraints on cooperation for both leader and potential partner might predict that the United States will be able to build larger coalitions than it otherwise might, since compensating any one partner will be cheaper and the military value of any one contribution will be greater. Such an insight, which might help address traditional questions about the formation of balancing coalitions (Christensen and Snyder 1990, Walt 1987), awaits further research, particularly on how potential balancers can use side payments to shape the terms of one another's cooperation.

Once inside a crisis, the logic of Chapter 4 suggests, perhaps reassuringly, that skittish members of American-led coalitions might serve a function similar to the one they served in the high-stakes environment of the Berlin Crisis: discouraging risky bluffing against a powerful opponent. The security environment in East Asia would be conducive to great power crises, as China seeks to redress territorial claims with neighbors like India and Russia, maritime claims against a host of regional powers from Japan to the Philippines to Vietnam, and the long-standing dispute over Taiwan's separation from the mainland. Many of these states will, and have already begun to, turn toward the United States to act as a balancer in the region, suggesting that the challenge of coalition-building and the management of cooperation will not be merely an artifact of unipolarity. Renewed great power competition might also mean that states choose not whether to cooperate or not, but which side to join, mirroring Italy's choice between rival coalitions in the First World War. Changing power realities might alter the expected course or outcome of crises, perhaps for the better in terms of escalation, but less so if a new great power rivalry will produce coalitions of less diverse preferences. Powerful but homogenous coalitions are both more prone to conflict *and* the provocation of balancing and counter-coalitions than are their more diverse counterparts. However, when states choose sides in the shadow of major wars, as they did twice in the twentieth century, military necessity's role in choosing partners may trump these potential second-order effects (cf. Riker 1962, Starr 1972), whether on signaling, durability, or balancing – suggesting that the problems associated with diverse coalitions may become greater when the potential stakes of the war itself grow larger.

Prediction of any sort is difficult, all the more so in the complex environment of the international system (see Braumoeller 2012, Jervis 1997). While I have not, in this section, predicted future power configurations, lines of battle, coalition makeup, or crisis outcomes, I hope to have provided a framework in which to understand coalitional politics for the United States if the world *does* change in the near future. Given its unique position atop the international power hierarchy, the United States will consistently find itself fighting abroad and relying on the assistance of distant countries to do so, regardless of how long it remains the sole superpower. Even if that changes, however, and it finds itself less often possessed of an overwhelming military advantage, we can simply "fit" the United States into the proper part of the parameter space in each of the theoretical models presented in this book and generate a (hopefully) useful set of expectations over what the future might hold in terms of military cooperation, war, and peace.

6.4 A COALITIONS RESEARCH AGENDA

This chapter has taken some preliminary steps toward extending the logic of military multilateralism in other directions. First, it leveraged the transactional theory of military cooperation to shed new light on the question of international cooperation in general, highlighting how the public or private nature of the gains from cooperation shapes both incentives to cooperate (or refuse) and the willingness to compensate (or not) that cooperation. Second, it made some informed conjectures about the differences between allied and nonallied coalitions, suggesting new lines of inquiry beyond the comparison of coalitions and singletons that animated most of the preceding discussion. Third, it showed how the models of previous chapters can help understand the foreign policies of specific countries – that is, the United States – when they engage in coalitional action. In closing, it is worth outlining some of the questions that remain unanswered, especially in terms of fleshing out the life cycle of coalitions from birth to death that Chapters 3–5 have only sketched in rough form.

Although it emerged as a secondary empirical implication in Chapter 5, the question of postwar coalitional durability is an obvious and potentially fruitful area for future research, especially as a transition to considerations of systemic and world-order politics. To the extent that similar dynamics inform coalitional durability after both small and large wars, the theory presented here provides a set of microfoundations linking the choice of coalition partner in systemic wars to the prospects for postwar peace among the remaining great powers; in starker terms, coalitional durability can often go on to mean systemic stability. Such choices – like the Western democracies' decision to tie their military fortunes to those of the Soviet Union in the Second World War – are often made as the result of sheer military exigency and the immediacy of present threats

(Riker 1962, Starr 1972). However strong their shared interest in fighting a common enemy, partners with widely divergent foreign policy preferences, particularly disagreements about the "distribution of ideology" (Braumoeller 2012, pp. 204–206), or relative shares of regime types in the international system, might win the war only to consign themselves to a dangerously unstable peace. Indeed, "[t]he difference between friend and foe can simply be a matter of time" (Plokhy 2010, p. xviii), and the empirical results over diverse preferences suggest that giving more states, or a more diverse set of states, a stake in the outcome of wars is not in itself a recipe for postwar stability.

Erstwhile partners may see cooperation break down over a number of issues related to the war – the creation of new postwar institutions, politics in vanquished state(s), the distribution of spoils, and honoring prior commitments of compensation. It is possible, however, that coalition-builders and their partners may be able to resolve such issues ahead of time by taking them off the table, choosing partners with whom they are less likely to disagree over how to manage the peace following whatever war they must first cooperate to win. When the choice of partner is less free, as it was for the Allies on Germany's frontiers in 1941, they may also structure cooperation in wartime so as to minimize future conflicts, perhaps by designating exclusive zones of military operation and control, which might reduce conflict over the spoils, or choosing particular compensation schemes. On the one hand, payments beforehand might encourage partners to defect once compensation is delivered, while on the other, compensation to be made afterwards might be inherently incredible.

Thus, another issue of immediate interest is the credibility of side payment schemes. In each preceding chapter, all promises of compensation are assumed to be credible; in other words, the lead state honors its promises by construction. Yet, as discussed briefly in Chapter 2, states are often worried over whether promises of compensation or cooperation will be granted. Lead states may refuse to make promised side payments after cooperation has been secured, just as nominal partners might refuse to cooperate if concessions were made up front. Alliance scholarship has examined the credibility of promises to cooperate (see Johnson and Leeds 2011, Leeds 2003a, Smith 1995), and the costs of failing to deliver cooperation undoubtedly play a role in coalitional behavior. The theory presented in this book suggests that the other problem – whether promised compensation is actually forthcoming – might illuminate both the formation of coalitions and the forms of compensation that lead states choose to promise. Italy's decision over choosing sides in the First World War is instructive again: the Central Powers offered concessions of Austrian territory that would be granted after the war, while the Allies promised that Italy could keep, essentially, whatever Austrian territory it conquered (Philpott 2014). While its postwar ambitions would be frustrated by its own failure to make significant headway against Austria, Italy would nonetheless take the latter offer, which was understandably viewed as far more credible than the former.[6] How, then, do states choose compensation schemes to maximize their credibility while, at the same time, securing cooperation at low cost? Further, how might the choice of compensation strategy – say, ex ante or ex post – affect crisis escalation and conflict expansion?

Domestic politics are also largely absent from the present account, and yet they undoubtedly shape many of the processes at the heart of intra-coalitional politics, from the distribution of the costs of war to ultimate ratification of the decision to enter coalitional conflicts. Democratic states require the support of the public before they join military coalitions, and to the extent that the public is hesitant or casualty-averse, they might require rather large premiums to offer their cooperation – just as Turkey did in negotiations over joining the Iraq War coalition in 2003. If democratic states drive harder bargains to join military coalitions, they might be more difficult partners to secure on average than nondemocratic states; those that join coalitions might, then, receive a larger share of the spoils than non-democracies. However, to the extent that democracies do not pursue spoils that can be treated as private rewards for supporters (Bueno de Mesquita et al. 2003, Goemans 2000), their concerns about public goods might shape both the types of coalitions they form *and* the extent to which collective action problems plague attempts to recruit them. Finally, when conflict outcomes affect leaders' political security (Bueno de Mesquita et al. 2003, Chiozza and Goemans 2004, Debs and Goemans 2010), the desire to demonstrate competence by successfully managing a coalition or securing a victory might make leaders more willing to compensate partners than those who do not share such concerns (see, e.g., Wolford and Ritter n.d.). Whether in terms of cost sensitivity, the executive's need to secure the support of legislators or the public, or political survival incentives, domestic politics offer a potentially fruitful area of research into the politics of military coalitions.

Next, cooperative problems within wartime coalitions have received extensive attention, especially following major coalitional efforts in Afghanistan and Iraq (e.g., Auerswald and Saideman 2014, Byman and Waxman 2002, Weitsman 2004), often in terms of the classic collective action problem (Olson 1965). However, the theory presented in this book abstracts away from such problems explicitly; states promise to cooperate in fighting a war and, if called upon to do so, follow through. We might imagine, however, that the terms of cooperation – whatever a lead state has to promise in return for a partner's cooperation – can have significant effects on the aims, duration, and outcome of the war, as well as the quality of cooperation between partners. For example, Saudi Arabia's unique ability to restrain American aims in 1991 almost certainly limited the aims and duration of the Gulf War, particularly relative

---

[6] In fact, Austrian leaders believed (their own dismal military performance to that point in the war notwithstanding) that whatever they might have to concede to Italy could be retaken by force at a later date (Philpott 2014).

to a case in which the United States would have had more freedom in choosing partners or an invasion route. Likewise, as the grim logic of attrition set in on the Western Front in the First World War, highlighting the need to beat the German army in the field to win the war, civilian leaders concerned over the distribution of war costs pulled Allied forces off the central front in 1917, lessening the attritional pressure on the German army and potentially lengthening the war (cf. Philpott 2014, Stevenson 2004). Further, promises of the post-victory spoils might encourage greater cooperation during the war if partners expect to control the land on which they sit at war's end, while ambiguous or incredible promises of compensation might impede cooperation. Recent theoretical models of war onset and termination typically examine only bilateral wars (e.g., Powell 2004, Slantchev 2003, Wagner 2000, Wolford, Reiter, and Carrubba 2011), but the approach here might suggest ways to embed a theory of coalitional cooperation into a theory of warfighting, which might yield new insights that strictly dyadic theoretical models cannot generate.

Finally, this book makes a larger point about the study of international conflict and cooperation, in that it highlights the importance of recognizing that the international system is not always best considered a collection of dyads. Chapters 3–5, in particular, show that how we choose to answer questions about cooperation, the probability of war, and the expansion of conflict – that is, whether we examine pairs of states or coalitions – can have a substantial impact on the answers we find. However, merely accounting for the multilateral nature of many conflicts empirically is insufficient without a strong theoretical foundation. The preliminary empirical work in Chapter 2, for example, found that coalitions are both more likely to see their crises escalate to war and expand to new participants than are states acting alone. Faced with these patterns, it can be tempting to draw the inference that coalitions simply have trouble signaling or credibly conveying either resolve or restraint, leading to war or conflict expansion, respectively (e.g., Byman and Waxman 2002, Christensen 2011, Lake 2010/2011, Vasquez 2009). However, the subsequent chapters showed that we should resist that temptation; the differences between coalitions and states acting alone cannot be reduced simply to differences in their military power. Coalitions are, quite simply, more (if sometimes less) than the sum of their parts. Uncovering precisely what makes them unique holds the promise of advancing still further our understanding of the ebb and flow of international war and peace.

Bibliography

Ai, Chunrong and Edward C. Norton. 2003. "Interaction Terms in Logit and Probit Models." *Economics Letters* 80(1):123–129.

Akcinaroglu, Seden. 2012. "Rebel Interdependencies and Civil War Outcomes." *Journal of Conflict Resolution* 56(5):879–903.

Alexander, Martin S. and William Philpott. 2002a. Introduction: Choppy Channel Waters – the Crests and Troughs of Anlgo-French Defence Relations between the Wars. In *Anglo-French Defence Relations between the Wars*, ed. Martin S. Alexander and William Philpott. New York: Palgrave Macmillan.

Alexander, Martin S. and William Philpott, eds. 2002b. *Anglo-French Defence Relations between the Wars*. New York: Palgrave Macmillan.

Altfeld, Michael D. and Bruce Bueno de Mesquita. 1979. "Choosing Sides in Wars." *International Studies Quarterly* 23(1):87–112.

Amin, Shahid M. 2010. *Pakistan's Foreign Policy: A Reappraisal*. 2nd ed. Oxford: Oxford University Press.

Angrist, Joshua D. and Jörn-Steffen Pischke. 2009. *Mostly Harmless Econometrics: An Empiricist's Companion*. Princeton: Princeton University Press.

Aono, Toshihiko. 2010. "'It Is Not Easy for the United States to Carry the Whole Load': Anglo-American Relations during the Berlin Crisis, 1961–1962." *Diplomatic History* 34(2):325–356.

Arena, Philip. n.d. "Costly Signaling, Resolve, and Martial Effectiveness." Typescript, University of Rochester.

Atkinson, Rick. 1993. *Crusade: The Untold Story of the Persian Gulf War*. New York: Houghton Mifflin.

Auerswald, David P. 2004. "Explaining Wars of Choice: An Integrated Decision Model of NATO Policy in Kosovo." *International Studies Quarterly* 48(3):631–662.

Auerswald, David P. and Stephen M. Saideman. 2014. *NATO in Afghanistan: Fighting Together, Fighting Alone*. Princeton: Princeton University Press.

Banks, Jeffrey D. 1990. "Equilibrium Behavior in Crisis Bargaining Games." *American Journal of Political Science* 34(3):599–614.

Baron, David P. and John A. Ferejohn. 1989. "Bargaining in Legislatures." *American Political Science Review* 83(4):1181–1206.

Bellamy, Alex J. 2000. "Lessons Unlearned: Why Coercive Diplomacy at Rambouillet Failed." *International Peacekeeping* 7(2):95–114.

Bennett, Andrew, Joseph Lepgold, and Danny Unger. 1994. "Burden-Sharing in the Persian Gulf War." *International Organization* 48(1):39–75.

Bennett, D. Scott. 1997. "Testing Alternative Models of Alliance Duration, 1816–1984." *American Journal of Political Science* 41(3):846–878.

Bennett, D. Scott and Allan C. Stam. 1996. "The Duration of Interstate Wars, 1816–1985." *American Political Science Review* 90(2):239–257.

Bennett, D. Scott and Allan C. Stam. 2004. *The Behavioral Origins of War.* Ann Arbor: University of Michigan Press.

Benson, Brett V. 2012. *Constructing International Security: Alliances, Deterrence, and Moral Hazard.* New York: Cambridge University Press.

Blinder, Alan S. 1973. "Wage Discrimination: Reduced Form and Structural Estimates." *Journal of Human Resources* 8(4):436–455.

Box-Steffensmeier, Janet M. and Bradford S. Jones. 2004. *Event History Modeling: A Guide for Social Scientists.* New York: Cambridge University Press.

Brambor, Thomas, William Roberts Clark, and Matt Golder. 2006. "Understanding Interaction Models: Improving Empirical Analyses." *Political Analysis* 14(1):63–82.

Braumoeller, Bear F. 2012. *The Great Powers and the International System: Systemic Theory in Empirical Perspective.* New York: Cambridge University Press.

Brecher, Michael and Jonathan Wilkenfeld. 1997. *A Study of Crisis.* Ann Arbor: University of Michigan Press.

Bueno de Mesquita, Bruce, Alastair Smith, Randolph M. Siverson, and James D. Morrow. 2003. *The Logic of Political Survival.* Cambridge, MA: MIT Press.

Bull, Hedley. 1977. *The Anarchical Society: A Study of Order in World Politics.* New York: Columbia University Press.

Bush, George H. W. and Brent Scowcroft. 1998. *A World Transformed.* New York: Knopf.

Bush, George W. 2002. "The President's State of the Union Address." Office of the Press Secretary, January 29.

Byman, Daniel and Matthew Waxman. 2002. *The Dynamics of Coercion.* New York: Cambridge University Press.

Carrubba, Clifford J. 2005. "Courts and Compliance in International Regulatory Regimes." *Journal of Politics* 67(3):669–689.

Carter, David B. 2010. "The Strategy of Territorial Conflict." *American Journal of Political Science* 54(4):969–987.

Chan, Steve. 2005. "Is There a Power Transition between the U.S. and China? The Different Faces of National Power." *Asian Survey* 45(5):687–701.

Chan, Steve. 2008. *China, the U.S., and the Power-Transition Theory: A Critique.* Oxford: Routledge.

Chapman, Terrence L. 2011. *Securing Approval: Domestic Politics and Multilateral Authorization for War.* Chicago: University of Chicago Press.

Chapman, Terrence L. and Dan Reiter. 2004. "The UN Security Council and the 'Rally Round-the-Flag' Effect." *Journal of Conflict Resolution* 48(6):886–909.

Chapman, Terrence L. and Scott Wolford. 2010. "International Organizations, Strategy, and Crisis Bargaining." *Journal of Politics* 72(1):227–242.

Chiba, Daina, Carla Martinez Machain, and William Reed. 2014. "Major Powers and Militarized Conflict." *Journal of Conflict Resolution* 58(6):976–1002.

Chiozza, Giacomo and Hein E. Goemans. 2004. "International Conflict and the Tenure of Leaders: Is War Still *Ex Post* Inefficient?" *American Journal of Political Science* 48(3):604–619.

Cho, In-Koo and David M. Kreps. 1987. "Signaling Games and Stable Equilibria." *Quarterly Journal of Economics* 102(2):179–221.

Choi, Ajin. 2004. "Democratic Synergy and Victory in War, 1816–1992." *International Studies Quarterly* 48(3):663–682.

Choi, Ajin. 2012. "Fighting to the Finish: Democracy and Commitment in Coalition War." *Security Studies* 21(4):624–653.

Christensen, Thomas J. 2011. *Worse Than a Monolith: Alliance Politics and Problems of Coercive Diplomacy in Asia*. Princeton: Princeton University Press.

Christensen, Thomas J. and Jack Snyder. 1990. "Chain Gangs and Passed Bucks: Predicting Alliance Patterns in Multipolarity." *International Organization* 44(2):137–138.

Clark, Christopher. 2012. *The Sleepwalkers: How Europe Went to War in 1914*. New York: HarperCollins.

Clark, Wesley K. 2001. *Waging Modern War*. New York: PublicAffairs.

CNN. 2003. "U.S. to move operations from Saudi base." http://edition.cnn.com/2003/WORLD/meast/04/29/sprj.irq.saudi.us/.

Colaresi, Michael, Karen Rasler, and William R. Thompson. 2007. *Strategic Rivalries in World Politics: Position, Space, and Conflict Escalation*. New York: Cambridge University Press.

Coll, Steve. 2004. *Ghost Wars: The Secret History of the CIA, Afghanistan and Bin Laden, from the Soviet Invasion to September 10, 2001*. New York: Penguin.

Corbetta, Renato and William J. Dixon. 2004. "Multilateralism, Major Powers, and Militarized Disputes." *Political Research Quarterly* 57(1):5–14.

COW. 2011. "Correlates of War State System Membership List, v2011." http://correlatesofwar.org.

Cremasco, Maurizio. 2000. Italy and the Management of International Crises. In *Alliance Politics, Kosovo, and NATO's War: Allied Force or Forced Allies?*, ed. Pierre Martin and Mark R. Brawley. New York: Palgrave.

Croco, Sarah E. and Tze Kwang Teo. 2005. "Assessing the Dyadic Approach to Interstate Conflict Processes: A.K.A. "Dangerous" Dyad-Years." *Conflict Management and Peace Science* 22(1):5–18.

Daalder, Ivo H. and Michael E. O'Hanlon. 2000. *Winning Ugly: NATO's War to Save Kosovo*. Washington, DC: The Brookings Institution.

Dallek, Robert. 2003. *An Unifinished Life: John F. Kennedy, 1917–1963*. New York: Little, Brown and Company.

Debs, Alexandre and H. E. Goemans. 2010. "Regime Type, the Fate of Leaders, and War." *American Political Science Review* 104(3):430–445.

Debs, Alexandre and Nuno P. Monteiro. 2014. "Known Unknowns: Power Shifts, Uncertainty, and War." *International Organization* 68(1):1–31.

Downs, George W., David M. Rocke, and Peter N. Barsoom. 1996. "Is the Good News about Compliance Good News about Cooperation?" *International Organization* 50(3):379–406.

Fairlie, Robert W. 2005. "An Extension of the Blinder-Oaxaca Decomposition Technique to Logit and Probit Models." *Journal of Economics and Social Measurement* 30(4):305–316.

Fang, Songying and Kristopher W. Ramsay. 2010. "Outside Options and Burden Sharing in Nonbinding Alliances." *Political Research Quarterly* 63(1):188–202.

Favretto, Katja. 2009. "Should Peacemakers Take Sides? Major Power Mediation, Coercion, and Bias." *American Political Science Review* 103(2):248–263.

Fearon, James D. 1994. "Domestic Political Audiences and the Escalation of International Disputes." *American Political Science Review* 88(3):577–692.

Fearon, James D. 1995. "Rationalist Explanations for War." *International Organization* 49(3):379–414.

Fearon, James D. 1997. "Signaling Foreign Policy Interests: Tying Hands versus Sinking Costs." *Journal of Conflict Resolution* 41(1):68–90.

Fortna, Virginia Page. 2003. "Scraps of Paper? Agreements and the Durability of Peace." *International Organization* 57(2):337–372.

Fowler, W.B. 1969. *British-American Relations 1917–1918: The Role of Sir William Wiseman*. Princeton: Princeton University Press.

Frank, Richard B. 2001. *Downfall: The End of the Imperial Japanese Empire*. New York: Penguin.

Frankfurt, Harry G. 2005. *On Bullshit*. Princeton: Princeton University Press.

Freedman, Lawrence. 2000. *Kennedy's Wars: Berlin, Cuba, Laos and Vietnam*. New York: Oxford University Press.

Freedman, Lawrence and Efraim Karsh. 1991. "How Kuwait Was Won: Strategy in the Gulf War." *International Security* 16(2):5–41.

Fromkin, David. 2004. *Europe's Last Summer: Who Started the Great War in 1914?* New York: Vintage.

Fudenberg, Drew and Jean Tirole. 1991. *Game Theory*. Cambridge, MA: MIT Press.

Fursenko, Aleksandr and Timothy Naftali. 2006. *Khrushchev's Cold War: The Inside Story of an American Adversary*. New York: W. W. Norton and Company.

Gamson, William A. 1961. "A Theory of Coalition Formation." *American Sociological Review* 26(3):373–382.

Gartner, Scott Sigmund and Randolph M. Siverson. 1996. "War Expanson and War Outcome." *Journal of Conflict Resolution* 40(1):4–15.

Gartzke, Erik and Kristian S. Gleditsch. 2004. "Why Democracies May Actually Be Less Reliable Allies." *American Journal of Political Science* 48(4):775–795.

Gaubatz, Kurt Taylor. 1996. "Democratic States and Commitment in International Relations." *International Organization* 50(1):109–139.

Ghosn, Faten, Glenn Palmer, and Stuart Bremer. 2004. "The MID3 Data Set, 1993–2001: Procedures, Coding Rules, and Description." *Conflict Management and Peace Science* 21:133–154.

Gibler, Douglas M. 2008. "The Costs of Reneging: Reputation and Alliance Formation." *Journal of Conflict Resolution* 52(3):426–454.

Gibler, Douglas M. and Meredith Reid Sarkees. 2004. "Measuring Alliances: The Correlates of War Formal Interstate Alliance Dataset, 1816-2000." *Journal of Peace Research* 41(2):211–222.

Gibler, Douglas M. and Scott Wolford. 2006. "Alliances, Then Democracy: An Examination of the Relationship Between Regime Type and Alliance Formation." *Journal of Conflict Resolution* 50(1):129–153.

Gibler, Douglas M. and Toby J. Rider. 2004. "Prior Commitments: Compatible Interests versus Capabilities in Alliance Behavior." *International Interactions* 30(4):309–329.

Gilligan, Michael, Leslie Johns, and B. Peter Rosendorff. 2010. "Strengthening International Courts and the Early Settlement of Disputes." *Journal of Conflict Resolution* 54(1):5–38.

Gleditsch, Kristian S. and Michael D. Ward. 2001. "Measuring Space: A Minimum Distance Database." *Journal of Peace Research* 38(6):739–758.

Goemans, H. E. 2000. *War and Punishment: The Causes of War Termination and the First World War*. Princeton: Princeton University Press.

Gordon, Michael R. and General Bernard E. Trainor. 2006. *Cobra II: The Inside Story of the Invasion and Occupation of Iraq*. New York: Pantheon.

Gowa, Joanne and Edward D. Mansfield. 1993. "Power Politics and International Trade." *American Political Science Review* 87(2):408–420.

Gowa, Joanne and Edward D. Mansfield. 2004. "Alliances, Imperfect Markets, and Major-Power Trade." *International Organization* 58(4):775–805.

Gulick, Edward Vose. 1967. *Europe's Classical Balance of Power*. New York: W. W. Norton and Company.

Haas, Mark L. 2007. "A Geriatric Peace? The Future of U.S. Power in a World of Aging Populations." *International Security* 32(1):112–147.

Haglund, David G. 2000. Allied Force or Forced Allies? The Allies' Perspective. In *Alliance Politics, Kosovo, and NATO's War: Allied Force or Forced Allies?*, ed. Pierre Martin and Mark R. Brawley. New York: Palgrave.

Hale, William M. 2007. *Turkey, the US and Iraq*. London: SAQI in Association with London Middle East Institute SOAS.

Hastings, Max. 2012. *Inferno: The World at War, 1939–1945*. New York: Vintage.

Hastings, Max. 2013. *Catastrophe 1914: Europe Goes to War*. New York: Knopf.

Heckman, James J. 1979. "Sample Selection Bias as a Specification Error." *Econometrica* 47(1):153–161.

Henriksen, Dag. 2007. *NATO's Gamble: Combining Diplomacy and Airpower in the Kosovo Crisis 1998–1999*. Annapolis: Naval Institute Press.

Hensel, Paul R., Sara McLaughlin Mitchell, Thomas E. Sowers II, and Clayton Thyne. 2008. "Bones of Contention: Comparing Territorial, Maritime, and River Issues." *Journal of Conflict Resolution* 52(1):117–143.

Herwig, Holger H. 2011. *The Marne, 1914: The Opening of World War I and the Battle That Changed the World*. New York: Random House.

Herz, John. 1951. *Political Realism and Political Idealism*. Chicago: University of Chicago Press.

Holsti, Kalevi J. 1991. *Peace and War: Armed Conflicts and International Order 1648–1989*. Cambridge: Cambridge University Press.

Horne, Alistair. 2006. *A Savage War of Peace: Algeria 1954–1962*. New York: The New York Review of Books.

Horne, Charles F. and Walter F. Austin, eds. 1930. *Source Records of the Great War*. Vol. II Indianapolis: The American Legion.

Huth, Paul and Bruce Russett. 1984. "What Makes Deterrence Work? Cases from 1900 to 1980." *World Politics* 36(4):496–526.

Huth, Paul, D. Scott Bennett, and Christopher Gelpi. 1992. "System Uncertainty, Risk Propensity, and International Conflict Among the Great Powers." *Journal of Conflict Resolution* 36(3):478–517.

Huth, Paul K. 1988. *Extended Deterrence and the Prevention of War*. New Haven: Yale University Press.

Huth, Paul K. 1997. "Reputations and Deterrence." *Security Studies* 7(1):72–99.

Ikeda, Maki and Atsushi Tago. 2014. "Winning Over Foreign Domestic Support for Use of Force: Power of Diplomatic and Operational Multilateralism." *International Relations of the Asia-Pacific* 14(2):303–324.

Ikenberry, G. John. 2001. *After Victory: Institutions, Strategic Restraint, and the Rebuilding of Order After Major Wars*. Princeton: Princeton University Press.

Iklé, Frank W. 1967. "The Triple Intervention: Japan's Lesson in the Diplomacy of Imperialism." *Monumenta Nipponica* 22(1/2):122–130.

Jervis, Robert. 1970. *The Logic of Images in International Relations*. Princeton: Princeton University Press.

Jervis, Robert. 1978. "Cooperation under the Security Dilemma." *World Politics* 30(2):167–214.

Jervis, Robert. 1997. *System Effects: Complexity in Political and Social Life*. Princeton: Princeton University Press.

Johnson, Jesse C. and Brett Ashley Leeds. 2011. "Defense Pacts: A Prescription for Peace?" *Foreign Policy Analysis* 7(1):45–65.

Jones, Seth G. 2010. *In the Graveyard of Empires: America's War in Afghanistan*. New York: W. W. Norton and Company.

Kalinovsky, Artemy M. 2011. *A Long Goobye: The Soviet Withdrawal from Afghanistan*. Cambridge, MA: Harvard University Press.

Karaca-Mandic, Pinar, Edward C. Norton, and Bryan Dowd. 2012. "Interaction Terms in Nonlinear Models." *Health Services Research* 47(1, Part I):255–274.

Keegan, John. 2000. *The First World War*. New York: Vintage.

Keegan, John. 2005. *The Iraq War*. New York: Vintage.

Kempe, Frederick. 2011. *Berlin 1961: Kennedy, Khrushchev, and the Most Dangerous Place on Earth*. New York: Putnam.

Kennan, George F. 1984. *American Diplomacy, Expanded Edition*. Chicago: University of Chicago Press.

Kennedy, Paul. 1989. *The Rise and Fall of the Great Powers*. New York: Vintage.

Keohane, Robert O. 1984. *After Hegemony: Cooperation and Discord in the World Political Economy*. Princeton: Princeton University Press.

Keohane, Robert O. 1990. "Multilateralism: An Agenda for Research." *International Journal* 45(4):731–764.

Keohane, Robert O. and Joseph S. Nye. 1977. *Power and Interdependence*. Boston: Little, Brown.

Kilcullen, David. 2006. "Counter-insurgency *Redux*." *Survival* 48(4):111–130.

Kimball, Anessa L. 2010. "Political Survival, Policy Distribution, and Alliance Formation." *Journal of Peace Research* 47(4):407–419.

Krauthammer, Charles. 1990. "The Unipolar Moment." *Foreign Affairs* 70(1):23–33.

Kreps, Sarah E. 2011. *Coalitions of Convenience: United States Military Interventions after the Cold War*. New York: Oxford University Press.

Kuniholm, Bruce R. 1991. "Turkey and the West." *Foreign Affairs* 70(2):34–48.

Kydd, Andrew H. 2005. *Trust and Mistrust in International Relations*. Princeton: Princeton University Press.

Lai, Brian. 2004. "The Effects of Different Types of Military Mobilization on the Outcome of International Crises." *Journal of Conflict Resolution* 48(2):211–229.

Lai, Brian and Dan Reiter. 2000. "Democracy, Political Similarity, and International Alliances, 1916–1992." *Journal of Conflict Resolution* 44(2):203–227.

Lai, Brian and Dan Reiter. 2005. "Rally 'Round the Union Jack? Public Opinion and the Use of Force in the United KIngdom, 1948–2001." *International Studies Quarterly* 49(2):255–272.

Lake, David A. 1999. *Entangling Relations: American Foreign Policy in Its Century.* Princeton: Princeton University Press.

Lake, David A. 2010/2011. "Two Cheers for Bargaining Theory: Assessing Rationalist Explanations of the Iraq War." *International Security* 35(3):7–52.

Leeds, Brett Ashley. 2003a. "Alliance Reliability in Times of War: Explaining State Decisions to Violate Treaties." *International Organization* 57(4):801–827.

Leeds, Brett Ashley. 2003b. "Do Alliances Deter Aggression? The Influence of Military Alliances on the Initiation of Militarized Interstate Disputes." *American Journal of Political Science* 47(3):427–439.

Leeds, Brett Ashley, Andrew G. Long, and Sara McLaughlin Mitchell. 2000. "Reevaluating Alliance Reliability: Specific Threats, Specific Promises." *Journal of Conflict Resolution* 44(5):686–699.

Leeds, Brett Ashley and Burcu Savun. 2007. "Terminating Alliances: Why Do States Abrogate Agreements?" *Journal of Politics* 69(4):1118–1132.

Leeds, Brett Ashley, Jeffrey M. Ritter, Sara McLaughlin Mitchell, and Andrew G. Long. 2002. "Alliance Treaty Obligations and Provisions, 1815–1944." *International Interactions* 28:237–260.

Leeds, Brett Ashley, Michaela Mattes, and Jeremy S. Vogel. 2009. "Interests, Institutions, and the Reliability of International Commitments." *American Journal of Political Science* 53(2):461–476.

Leetaru, Kalev and Scott Althaus. 2009. "Airbrushing History, American Style: The Mutability of Government Documents in the Digital Era." *D-LIB Magazine* 15(1–2) www.dlib.org/dlib/january09/01inbrief.html.

Leventoglu, Bahar and Ahmer Tarar. 2008. "Does Private Information Lead to Delay or War in Crisis Bargaining?" *International Studies Quarterly* 52(3):533–553.

Lo, Nigel, Barry Hashimoto, and Dan Reiter. 2008. "Ensuring Peace: Foreign-Imposed Regime Change and Postwar Peace Duration, 1914–2001." *International Organization* 62(4):717–736.

Mansfield, Edward D. and Rachel Bronson. 1997. "Alliances, Preferential Trading Agreements, and International Trade." *American Political Science Review* 91(1):94–107.

Marshall, Monty G. and Keith Jaggers. 2009. "Polity IV Project: Political Regime Characteristics and Transitions, 1800–2007." Polity IV dataset version 2007. www .systemicpeace.org/polity4.

Martin, Pierre and Mark R. Brawley, eds. 2000. *Alliance Politics, Kosovo, and NATO's War: Allied Force or Forced Allies?* New York: Palgrave.

Mattes, Michaela and Greg Vonnahme. 2010. "Contracting for Peace: Do Nonaggression Pacts Reduce Conflict?" *Journal of Politics* 72(4):925–938.

May, Ernest R. 2000. *Strange Victory: Hitler's Conquest of France.* New York: Hill & Wang.

Mearsheimer, John J. 1995. "Back to the Future: Instability in Europe after the Cold War." *International Security* 15(1):5–56.

Mearsheimer, John J. 2001. *The Tragedy of Great Power Politics*. New York: W. W. Norton and Company.

Meirowitz, Adam and Anne E. Sartori. 2008. "Strategic Uncertainty as a Cause of War." *Quarterly Journal of Political Science* 3(4):327–352.

Miers, Anne C. and T. Clifton Morgan. 2002. "Multilateral Sanctions and Foreign Policy Success: Can Too Many Cooks Spoil the Broth?" *International Interactions* 28(1):117–136.

Migdalovitz, Carol. 2003. *Iraq: Turkey, the Deployment of U.S. Forces, and Related Issues*. Washington, DC: Congressional Research Service.

Morey, Daniel S. 2011. "Military Coalitions and the Outcome of Interstate Wars." Presented at the 2011 Annual Meeting of the American Political Science Association, Seattle, WA.

Morgenthau, Hans Joachim. 1967. *Politics Among Nations*. 4th ed. New York: Knopf.

Morrow, James D. 1989. "Capabilities, Uncertainty, and Resolve: A Limited Information Model of Crisis Bargaining." *American Journal of Political Science* 33(4):941–972.

Morrow, James D. 1991. "Alliances and Asymmetry." *American Journal of Political Science* 35(4):904–933.

Morrow, James D. 1993. "Arms versus Allies: Trade-Offs in the Search for Security." *International Organization* 47(2):207–233.

Morrow, James D. 1994. "Alliances, Credibility, and Peacetime Costs." *Journal of Conflict Resolution* 38(2):270–297.

Morrow, James D. 2000. "Alliances: Why Write Them Down?" *Annual Review of Political Science* 3:63–83.

Mousseau, Michael. 1997. "Democracy and Militarized Interstate Collaboration." *Journal of Peace Research* 34(1):73–87.

Myerson, Roger B. 1992. "On the Value of Game Theory in Social Science." *Rationality and Society* 4(1):62–73.

Norton, Edward C., Hua Wang and Chunrong Ai. 2004. "Computing Interaction Effects and Standard Errors in Logit and Probit Models." *The Stata Journal* 4(2):154–167.

Nye, Joseph S. 2002. *The Paradox of American Power: Why the World's Only Superpower Can't Go It Alone*. New York: Oxford University Press.

Oaxaca, Ronald L. 1973. "Male-Female Differentials in Urban Labor Markets." *International Economic Review* 14(3):693–709.

Olson, Mancur. 1965. *The Logic of Collective Action: Public Goods and the Theory of Groups*. Cambridge, MA: Harvard University Press.

Olson, Mancur and Richard Zeckhauser. 1966. "An Economic Theory of Alliances." *Review of Economics and Statistics* 48(3):266–279.

Önis, Ziya and Suhnaz Yilmaz. 2005. "The Turkey-EU-US Triangle in Perspective: Transformation or Continuity?" *Middle East Journal* 59(2):265–284.

Oye, Kenneth A., ed. 1986. *Cooperation Under Anarchy*. Princeton: Princeton University Press.

Paine, S. C. M. 2012. *The Wars for Asia, 1911–1949*. New York: Cambridge University Press.

Palmer, Glenn and T. Clifton Morgan. 2010. *A Theory of Foreign Policy*. Princeton: Princeton University Press.

Papayoanou, Paul A. 1997. "Intra-Alliance Bargaining and U.S. Bosnia Policy." *Journal of Conflict Resolution* 41(1):91–116.

Philpott, William. 2009. *Three Armies on the Somme: The First Battle of the Twentieth Century*. New York: Vintage.

Philpott, William. 2011. "Review of Roy A. Prete. *Strategy and Command: The Anglo-French Coalition on the Western Front, 1914*. Montreal: Queen's University Press, 2009." *Journal of British Studies* 50(1):235–235.

Philpott, William. 2014. *War of Attrition: Fighting the First World War*. New York: The Overlook Press.

Pilster, Ulrich. 2011. "Are Democracies the Better Allies? The Impact of Regime Type on Mlitary Coalition Operations." *International Interactions* 37(1):55–85.

Plokhy, S.M. 2010. *Yalta: The Price of Peace*. New York: Penguin.

Poast, Paul. 2010. "(Mis)Using Dyadic Data to Analyze Multilateral Events." *Political Analysis* 18(4):403–425.

Poast, Paul. 2012. "Does Issue Linkage Work? Evidence from European Alliance Negotiations, 1860 to 1945." *International Organization* 66(2):277–310.

Powell, Robert. 1999. *In the Shadow of Power: States and Strategies in International Politics*. Princeton: Princeton University Press.

Powell, Robert. 2004. "Bargaining and Learning While Fighting." *American Journal of Political Science* 48(2):344–361.

Pressman, Jeremy. 2008. *Warring Friends: Alliance Restraint in International Politics*. Ithaca: Cornell University Press.

Primo, David M. and Kevin A. Clarke. 2012. *A Model Discipline: Political Science and the Logic of Representations*. New York: Oxford University Press.

Prost, Antoine. 2014. The Dead. In *The Cambridge History of the First World War*, ed. Jay Winter. Cambridge: Cambridge University Press.

Purrington, Courtney. 1992. "Tokyo's Policy Responses During the Gulf War and the Impact of the 'Iraqi Shock' on Japan." *Pacific Affairs* 65(2):161–181.

Reed, William. 2000. "A Unified Statistical Model of Conflict Onset and Escalation." *American Journal of Political Science* 44(1):84–93.

Reed, William. 2003. "Information, Power, and War." *American Political Science Review* 97(4):633–641.

Reed, William and Daina Chiba. 2010. "Decomposing the Relationship between Contiguity and Militarized Conflict." *American Journal of Political Science* 54(1): 61–73.

Reed, William, David H. Clark, Timothy Nordstrom, and Wonjae Hwang. 2008. "War, Power, and Bargaining." *Journal of Politics* 70(4):1203–1216.

Reiter, Dan. 2009. *How Wars End*. Princeton: Princeton University Press.

Reiter, Dan and Allan C. Stam. 2002. *Democracies at War*. Princeton: Princeton University Press.

Remnick, David. 2003. "War Without End?" *The New Yorker*, April 23.

Resnick, Evan N. 2010/2011. "Strange Bedfellows: US Bargaining Behavior with Allies of Convenience." *International Security* 35(3):144–184.

Richardson, Louise. 2000. A Force for Good in the World? Britain's Role in the Kosovo Crisis. In *Alliance Politics, Kosovo, and NATO's War: Allied Force or Forced Allies?*, ed. Pierre Martin and Mark R. Brawley. New York: Palgrave.

Riker, William H. 1962. *The Theory of Political Coalitions*. New Haven: Yale University Press.

Robins, Philip. 2003. "Confusion at Home, Confusion Abroad: Turkey between Copenhagen and Iraq." *International Affairs* 79(3):547–566.

Rosendorff, B. Peter. 2005. "Stability and Rigidity: Politics and Design of the WTO's Dispute Settlement Procedure." *American Political Science Review* 99(3):389–400.

Rudolf, Peter. 2000. Germany and the Kosovo Conflict. In *Alliance Politics, Kosovo, and NATO's War: Allied Force or Forced Allies?*, ed. Pierre Martin and Mark R. Brawley. New York: Palgrave.

Russett, Bruce, John R. Oneal, and David R. Davis. 1998. "The Third Leg of the Kantian Tripod for Peace: International Organizations and Militarized Disputes, 1950–1985." *International Organization* 52(3):441–467.

Russett, Bruce M. 1963. "The Calculus of Deterrence." *Journal of Conflict Resolution* 7(2):97–109.

Sandler, Todd. 1992. *Collective Action: Theory and Applications*. Ann Arbor: University of Michigan Press.

Sandler, Todd and Keith Hartley. 2001. "Economics of Alliances: The Lessons for Collective Action." *Journal of Economic Literature* 39(3):869–896.

Sarkees, Meredith Reid and Frank Wayman. 2010. *Resort to War: 1816–2007*. Washington, DC: CQ Press.

Sayari, Sabri. 1992. "Turkey: The Changing European Security Environment and the Gulf Crisis." *Middle East Journal* 46(1):9–21.

Schelling, Thomas C. 1966. *Arms and Influence*. New Haven: Yale University Press.

Schultz, Kenneth A. 1998. "Domestic Opposition and Signaling in International Crises." *American Political Science Review* 92(4):829–844.

Schultz, Kenneth A. 2001. "Looking for Audience Costs." *Journal of Conflict Resolution* 45(1):32–60.

Schultz, Kenneth A. and Barry R. Weingast. 2003. "The Democratic Advantage: Institutional Foundations of Financial Power and International Competition." *International Organization* 57(1):3–42.

Schweller, Randall L. 1994. "Bandwagoning for Profit: Bringing the Revisionist State Back In." *International Security* 19(1):72–107.

Sechser, Todd S. 2010. "Goliath's Curse: Coercive Threats and Asymmetric Power." *International Organization* 64(4):627–660.

Senese, Paul D. and John A. Vasquez. 2008. *The Steps to War: An Empirical Study*. Princeton: Princeton University Press.

Shirkey, Zachary C. 2012. "When and How Many: The Effects of Third Party Joining on Casualties and Duration in Interstate Wars." *Journal of Peace Research* 49(2):321–334.

Singer, David J., Stuart Bremer, and John Stuckey. 1972. Capability Distribution, Uncertainty, and Major Power War, 1820–1965. In *Peace, War, and Numbers*, ed. Bruce Rusett. Beverly Hills: Sage.

Singer, J. David. 1987. "Reconstructing the Correlates of War Dataset on Material Capabilities of States, 1816–1985." *International Interactions* 14(2):115–132.

Siverson, Randolph M. and Harvey Starr. 1990. "Opportunity, Willingness, and the Diffusion of War." *American Political Science Review* 84(1):47–67.

Siverson, Randolph M. and Juliann Emmons. 1991. "Birds of a Feather: Democratic Political Systems and Alliance Choices in the Twentieth Century." *Journal of Conflict Resolution* 35(2):285–306.

Slantchev, Branislav. 2003. "The Principle of Convergence in Wartime Negotiations." *American Political Science Review* 47(4):621–632.

Slantchev, Branislav. 2004. "How Initiators End Their Wars: The Duration of Warfare and the Terms of Peace." *American Journal of Political Science* 48(4):813–829.

Slantchev, Branislav. 2005. "Military Coercion in Interstate Crises." *American Political Science Review* 99(4):533–547.

Slantchev, Branislav. 2010. "Feigning Weakness." *International Organization* 64(3):357–388.

Slantchev, Branislav. 2011. *Military Threats: The Costs of Coercion and the Price of Peace.* Cambridge: Cambridge University Press.

Smith, Alastair. 1995. "Alliance Formation and War." *International Studies Quarterly* 39(4):405–425.

Smith, Alastair. 1998. "International Crises and Domestic Politics." *American Political Science Review* 92(3):623–638.

Snyder, Glenn H. 1984. "The Security Dilemma in Alliance Politics." *World Politics* 36(4):461–495.

Snyder, Glenn H. 1991. "Alliances, Balance, and Stability." *International Organization* 45(1):121–142.

Snyder, Glenn H. 1997. *Alliance Politics.* Ithaca: Cornell University Press.

Sobek, David and Joe Clare. 2013. "Me, Myself, and Allies: Understanding the External Sources of Power." *Journal of Peace Research* 50(4):469–478.

Spence, Michael. 1973. "Job Market Signaling." *Quarterly Journal of Economics* 87(3):355–374.

Starr, Harvey. 1972. *War Coalitions: The Distributions of Payoffs and Losses.* Lexington: Lexington Books.

Stevenson, David. 1997. "Militarization and Diplomacy in Europe before 1914." *International Security* 22(1):125–161.

Stevenson, David. 2004. *Cataclysm: The First World War as Political Tragedy.* New York: Basic Books.

Stinnett, Douglas M., Jaroslav Tir, Philip Schafer, Paul F. Diehl, and Charles Gochman. 2002. "The Correlates of War Project Direct Contiguity Data, Version 3." *Conflict Management and Peace Science* 19(2):58–66.

Strezhnev, Anton and Erik Voeten. 2013. "United Nations General Assembly Voting Data." http://hdl.handle.net/1902.1/12379.

Stueck, William. 1995. *The Korean War: An International History.* Princeton: Princeton University Press.

Stueck, William. 2004. *Rethinking the Korean War: A New Diplomatic and Strategic History.* Princeton: Princeton University Press.

Tago, Atsushi. 2007. "Why Do States Join US-led Military Coalitions?: The Compulsion of the Coalition's Missions and Legitimacy." *International Relations of the Asia-Pacific* 7(2):179–202.

Tago, Atsushi. 2009. "When Are Democratic Friends Unreliable? The Unilateral Withdrawal of Troops from the 'Coalition of the Willing'." *Journal of Peace Research* 46(2):219–234.

Tarar, Ahmer. 2013. "Military Mobilization and Commitment Problems." *International Interactions* 39(3):343–366.

Tarar, Ahmer and Bahar Leventoğlu. 2013. "Limited Audience Costs in International Crises." *Journal of Conflict Resolution* 57(6):1065–1089.

Thompson, Alexander. 2006. "Coercion Through IOs: The Security Council and the Logic of Information Transmission." *International Organization* 60(1):1–34.

Tobin, James. 1958. "Estimation of Relationships for Limited Dependent Variables." *Econometrica* 26(1):24–36.

Trager, Robert F. 2010. "Diplomatic Calculus in Anarchy: How Communication Matters." *American Political Science Review* 104(2):347–368.

Vasquez, John A. 2009. *The War Puzzle Revisited*. Cambridge: Cambridge University Press.

Voeten, Erik. 2001. "Outside Options and the Logic of Security Council Action." *American Political Science Review* 95(4):845–858.

Voeten, Erik. 2005. "The Political Origins of the UN Security Council's Ability to Legitimize the Use of Force." *International Organization* 59:527–557.

Vucetic, Srdjan. 2011. "Bound to Follow? The Anglosphere and US-Led Coalitions of the Willing, 1950–2001." *European Journal of International Relations* 17(1):27–49.

Wagner, R. Harrison. 2000. "Bargaining and War." *American Journal of Political Science* 44(3):469–484.

Wagner, R. Harrison. 2007. *War and the State: The Theory of International Politics*. Ann Arbor: The University of Michigan Press.

Wallander, Celeste A. 2000. "Institutional Assets and Adaptability: NATO After the Cold War." *International Organization* 54(4):705–735.

Walt, Stephen M. 1987. *The Origins of Alliances*. Ithaca: Cornell University Press.

Waltz, Kenneth. 1979. *Theory of International Politics*. Reading: Addison-Wesley.

Ward, Michael D. 1982. *Research Gaps in Alliance Dynamics*. Vol. 19 of *Monograph Series in World Affairs*. Denver: Graduate School of International Studies, University of Denver.

Weinbaum, Marvin G. 1991a. "Pakistan and Afghanistan: The Strategic Relationship." *Asian Survey* 31(6):496–511.

Weinbaum, Marvin G. 1991b. "War and Peace in Afghanistan: The Pakistani Role." *Middle East Journal* 45(1):71–85.

Weitsman, Patricia A. 2004. *Dangerous Alliances: Proponents of Peace, Weapons of War*. Stanford: Stanford University Press.

Werner, Suzanne. 1999. "The Precarious Nature of Peace: Resolving the Issues, Enforcing the Settlement, and Renegotiating the Terms." *American Journal of Political Science* 43(3):912–934.

Werner, Suzanne. 2000. "Deterring Intervention: The Stakes of War and Third-Party Involvement." *American Journal of Political Science* 44(4):720–732.

Werner, Suzanne and Amy Yuen. 2005. "Making and Keeping Peace." *International Organization* 59(2):261–292.

Wilkenfeld, Jonathan and Michael Brecher. 2010. *International Crisis Behavior Project*. v.10 Ann Arbor: ICPSR Study #9286. http://www.cidcm.umd.edu/icb/data.

Winter, Jay, ed. 2014. *The Cambridge History of the First World War*. Cambridge: Cambridge University Press.

Wolford, Scott. 2007. "The Turnover Trap: New Leaders, Reputation, and International Conflict." *American Journal of Political Science* 51(4):772–788.

Wolford, Scott. 2013. "Theorizing Multilateralism in a Time of Unipolarity." *International Studies Review* 15(2):295–297.

Wolford, Scott. 2014a. "Power, Preferences, and Balancing: The Durability of Coalitions and the Expansion of Conflict." *International Studies Quarterly* 58(1):146–157.

Wolford, Scott. 2014b. "Showing Restraint, Signaling Resolve: Coalitions, Cooperation, and Crisis Bargaining." *American Journal of Political Science* 58(1):144–156.

Wolford, Scott, Dan Reiter, and Clifford J. Carrubba. 2011. "Information, Commitment, and War." *Journal of Conflict Resolution* 55(4):556–579.

Wolford, Scott and Emily Hencken Ritter. n.d. "National Leaders, Political Security, and Military Coalitions." Typescript, University of Texas and University of California, Merced.

Xiang, Lanxin. 2003. *The Origins of the Boxer War: A Multinational Study.* London: RoutledgeCurzon.

Yuen, Amy. 2009. "Target Concessions in the Shadow of Intervention." *Journal of Conflict Resolution* 53(5):727–744.

Zinnes, Dina A. 1967. "An Analytical Study of the Balance of Power Theories." *Journal of Peace Research* 4(3):270–287.

Index

CPSIA information can be obtained at www.ICGtesting.com
Printed in the USA
LVOW10s1150030916

503076LV00002B/261/P